Quick Reference for the
Civil Engineering PE Exam

Third Edition

Michael R. Lindeburg, PE

Professional Publications, Inc. • Belmont, CA

QUICK REFERENCE FOR THE CIVIL ENGINEERING PE EXAM
Third Edition

Current printing of this edition: 3

Printing History

edition number	printing number	update
3	1	Based on 8th edition of *Civil Engineering Reference Manual for the PE Exam*. Copyright updated.
3	2	Minor corrections.
3	3	Minor corrections. Synchronized with the 9th edition of *Civil Engineering Reference Manual for the PE Exam*.

Printed in the United States of America

Professional Publications, Inc.
1250 Fifth Avenue, Belmont, CA 94002
(650) 593-9119
www.ppi2pass.com

Library of Congress Cataloging-in-Publication Data
Lindeburg, Michael R.
    Quick reference for the civil engineering PE exam / Michael R. Lindeburg.--3rd ed.
        p. cm.
    ISBN 1-888577-76-2
    1. Civil engineering--United States--Examinations--Study guides. 2. Civil engineering--Handbooks, manuals, etc. I. Title.

TA159.L58 2003
624′.076--dc21
                                              2003046601

Table of Contents

How to Use This Book

This book (and others in the Quick Reference series) was developed to help you minimize problem-solving time on the PE exam. *Quick Reference* is a consolidation of the most useful equations in the *Civil Engineering Reference Manual*. Using *Quick Reference*, you won't need to wade through pages of descriptive text when all you actually need is a quick look at the formulas to remind you of your next solution step.

The idea is this: you study for the exam using (primarily) your *Reference Manual*, and you take the exam using (primarily) your *Quick Reference*.

Once you've studied and mastered the theory behind an exam topic, you're ready to tackle practice problems. Save time by using this *Quick Reference* right from the start, as you begin solving problems. This book follows the same order and uses the same nomenclature as the *Civil Engineering Reference Manual*. Once you become familiar with the sequencing of subjects in the *Reference Manual*, you'll be at home with *Quick Reference*. As you progress in your understanding of topics, you'll find you can rely more and more on *Quick Reference* for rapid retrieval of equations—without needing to refer back to the *Reference Manual*.

There are also times during problem-solving when you do need access to two kinds of information simultaneously—formulas and data, or formulas and nomenclature, or formulas and theory. We've all experienced the frustration of having to work problems using a spare pencil, a calculator, and the water-bill envelope to serve as page markers in a single book. *Quick Reference* provides a convenient way to keep the equations you need in front of you, even as you may be flip-flopping back and forth between theory and data in your other references.

Quick Reference also provides a convenient place for neatnik engineers to add their own comments and reminders to equations, without having to mess up their primary references. We expect you to throw this book away after the exam, so go ahead and write in it.

Once you start incorporating *Quick Reference* into your problem-solving routine, we predict you won't want to return to the one-book approach. *Quick Reference* will save you precious time—and that's how to use this book.

Codes Used to Prepare This Book

The information that was used to write and update this book was the most current at the time. However, as with engineering practice itself, the PE examination is not always based on the most current codes or cutting-edge technology. Similarly, codes, standards, and regulations adopted by state and local agencies often lag issuance by several years. Thus, the codes that are current, used by you in practice, and tested on the exam can all be different.

PPI lists on its website the dates and editions of the codes, standards, and regulations on which NCEES has announced the PE exams are based. It is your responsibility to find out which codes will be tested on your exam. In the meantime, here are the codes that have been incorporated into this edition.[1]

STRUCTURAL DESIGN STANDARDS

ACI 318: *Building Code Requirements for Structural Concrete*, 1999, American Concrete Institute, Farmington Hills, MI

ACI 530: *Building Code Requirements for Masonry Structures*, 1999, and ACI 530.1: *Specifications for Masonry Structures*, 1999, American Concrete Institute, Detroit, MI

AISC/ASD: *Manual of Steel Construction, Allowable Stress Design*, Ninth ed., 1989, American Institute of Steel Construction, Inc., Chicago, IL

AISC/LRFD: *Manual of Steel Construction Load and Resistance Factor Design*, Second ed., 1994, American Institute of Steel Construction, Inc., Chicago, IL

PCI: *PCI Design Handbook*, Fifth ed., 1999, Precast/Prestressed Concrete Institute, Chicago, IL

UBC: *Uniform Building Code*, 1997 ed., International Conference of Building Officials, Whittier, CA

TRANSPORTATION DESIGN STANDARDS

AASHTO: *A Policy on Geometric Design of Highways and Streets*, Fourth ed., 2001, American Association of State Highway & Transportation Officials, Washington, DC

AI: *The Asphalt Handbook* (MS-4), 1989, Asphalt Institute, College Park, MD

HCM: *Highway Capacity Manual*, Fourth ed., U.S. Customary version, 2000, Transportation Research Board, National Research Council, Washington, DC

MUTCD: *Manual on Uniform Traffic Control Devices*, 2000, U.S. Dept. of Transportation, Federal Highway Administration, Washington, DC

PCA: *Design and Control of Concrete Mixtures*, Fourteenth ed., 2002, Portland Cement Association, Skokie, IL

---

[1]This is a list of the codes used to prepare *this* book. The Introduction contains a complete list of the references that you should consider bringing to the exam.

For Instant Recall

Fundamental and Physical Constants

quantity	symbol	English	SI
Density			
air [STP] [32°F (0°C)]		0.0805 lbm/ft^3	1.29 kg/m^3
air [70°F (20°C), 1 atm]		0.0749 lbm/ft^3	1.20 kg/m^3
earth [mean]		345 lbm/ft^3	5520 kg/m^3
mercury		849 lbm/ft^3	1.360×10^4 kg/m^3
sea water		64.0 lbm/ft^3	1025 kg/m^3
water [mean]		62.4 lbm/ft^3	1000 kg/m^3
Specific Gravity			
mercury		13.6	
water		1.0	
Gravitational Acceleration			
earth [mean]	g	32.174 (32.2) ft/sec^2	9.8067 (9.81) m/s^2
Pressure, atmospheric		14.696 (14.7) lbf/in^2	1.0133×10^5 Pa
Temperature, standard		32°F (492°R)	0°C (273K)
Fundamental Constants			
Avogadro's number	N_A		6.022×10^{23} mol^{-1}
gravitational constant	g_c	32.174 lbm-ft/lbf-sec^2	
specific gas constant, air	R	53.3 ft-lbf/lbm-°R	287 J/kg·K
specific gas constant, methane	R_{CH_4}	96.32 ft-lbf/lbm-°R	518.3 J/kg·K
universal gas constant	R^*	1545 ft-lbf/lbmol-°R	8314 J/kmol·K
	R^*	1.986 BTU/lbmol-°R	0.08206 atm·L/mol·K
Molecular Weight			
air		29	
carbon		12	
carbon dioxide		44	
helium		4	
hydrogen		2	
methane		16	
nitrogen		28	
oxygen		32	
Steel			
modulus of elasticity		29×10^7 psi	
modulus of shear		1.2×10^7 psi	
Poisson's ratio		0.3	

Temperature Conversions

$$°F = 32 + \tfrac{9}{5}°C$$

$$°C = \tfrac{5}{9}(°F - 32)$$

$$°R = °F + 460$$

$$K = °C + 273$$

$$\Delta°R = \tfrac{9}{5}\Delta K$$

$$\Delta K = \tfrac{5}{9}\Delta°R$$

SI Prefixes

symbol	prefix	value
a	atto	10^{-18}
f	femto	10^{-15}
p	pico	10^{-12}
n	nano	10^{-9}
μ	micro	10^{-6}
m	milli	10^{-3}
c	centi	10^{-2}
d	deci	10^{-1}
da	deka	10
h	hecto	10^2
k	kilo	10^3
M	mega	10^6
G	giga	10^9
T	tera	10^{12}
P	peta	10^{15}
E	exa	10^{18}

Engineering Conversions

Unless noted otherwise, atmospheres are standard; Btus are IT (international table); calories are gram-calories; gallons are U.S. liquid; miles are statute; pounds-mass are avoirdupois; chains are surveyors'.

multiply	by	to obtain
ac	10.0	chain^2
ac	43,560	ft^2
ac	0.40469	hectare
ac	4046.87	m^2
ac	1/640	mi^2
ac-ft	43,560	ft^3
ac-ft	0.01875	m^2-in
ac-ft	1233.5	m^3
ac-ft	325,851	gal
ac-ft/day	0.50416	ft^3/s
ac-ft/mi^2	0.01875	in (runoff)
ac-in	102,790	L
ac-in/hr	1.0083	ft^3/s
angstrom	1.0×10^{-10}	m
atm	1.01325	bar
atm	76.0	cm Hg
atm	33.90	ft water
atm	29.921	in Hg
atm	14.696	lbf/in^2
atm	101.33	kPa
atm	1.0133×10^5	Pa
bar	0.9869	atm
bar	1.0×10^5	Pa
Btu	778.169	ft-lbf
Btu	1055.056	J
Btu	2.928×10^{-4}	kW-hr
Btu	1.0×10^{-5}	therm
Btu/hr	0.21611	ft-lbf/s
Btu/hr	3.929×10^{-4}	hp
Btu/hr	0.29307	W
Btu/lbm	2.3260	kJ/kg
Btu/lbm-°R	4.1868	kJ/kg·K
Btu/s	778.26	ft-lbf/s
Btu/s	46,680	ft-lbf/min
Btu/s	1.4148	hp
Btu/s	1.0545	kW
cal	3.968×10^{-3}	Btu
cal	4.1868	J
chain	66.0	ft
chain	1/80	mi (statute)
chain	4.0	rod
chain	22	yd
chain^2	1/10	ac
cm	0.03281	ft
cm	0.39370	in
cm/s	1.9686	ft/min
cm^2	0.1550	in^2
cm^3	2.6417×10^{-4}	gal
cm^3	0.0010	L
day (mean solar)	86,400	s
day (sidereal)	86,164.09	s
darcy	1.0623×10^{-11}	ft^2
degree (angular)	1/0.9	grad
degree (angular)	$2\pi/360$	radian
degree (angular)	17.778	mil
dyne	1.0×10^{-5}	N
eV	1.6022×10^{-19}	J
ft	1/66	chain
ft	0.3048	m
ft	1/5280	mi
ft	0.06060	rod
ft of water	0.43328	lbf/in^2
ft-kips	1.358	kN·m
ft-lbf (energy)	1.2851×10^{-3}	Btu
ft-lbf (energy)	1.3558	J
ft-lbf (energy)	3.766×10^{-7}	kW·h
ft-lbf (torque)	1.3558	N·m
ft-lbf/min	1/46,680	Btu/s
ft-lbf/min	1/33,000	hp
ft-lbf/min	2.2598×10^{-5}	kW

multiply	by	to obtain
ft-lbf/s	1.2849×10^{-3}	Btu/s
ft-lbf/s	1/550	hp
ft-lbf/s	1.3558×10^{-3}	kW
ft/min	0.50798	cm/s
ft/s	0.68180	mi/hr
ft^2	1/43,560	ac
ft^2	9.4135×10^{10}	darcy
ft^2	0.09290	m^2
ft^3	1/43,560	ac-ft
ft^3	1728.0	in^3
ft^3	7.4805	gal
ft^3	28.320	L
ft^3	0.03704	yd^3
ft^3/day	0.005195	gal/min
ft^3/mi^2	4.3×10^{-7}	in runoff
ft^3/min	10,772	gal/day
ft^3/s	1.9835	ac-ft/day
ft^3/s	0.99177	ac-in/hr
ft^3/s	646,317	gal/day
ft^3/s	448.83	gal/min
ft^3/s	1/26.88	m^2-in/day
ft^3/s	0.64632	MGD
g	0.03527	oz
g	0.002205	lbm
g/cm^3	1000.0	kg/m^3
g/cm^3	62.428	lbm/ft^3
gal	1/325,851	ac-ft
gal	3785.4	cm^3
gal	0.13368	ft^3
gal (Imperial)	1.2	gal (U.S.)
gal (U.S.)	0.8327	gal (Imperial)
gal	3.7854	L
gal	3.7854×10^{-3}	m^3
gal	1.0×10^{-6}	MG
gal	0.00495	yd^3
gal (of water)	8.34	lbm (of water)
gal/ac-day	0.04356	MG/ft^2-day
gal/day	9.283×10^{-5}	ft^3/min
gal/day	1.5472×10^{-6}	ft^3/s
gal/day	1/1440	gal/min
gal/day	1.0×10^{-6}	MG/day
gal/day-ft	0.01242	m^3/m·day
gal/day-ft^2	0.04075	m^3/m^2·day
gal/day-ft^2	1.0	Meinzer unit
gal/day-ft^2	0.04355	MGD/ac (mgad)
gal/min	0.002228	ft^3/s
gal/min	192.50	ft^3/day
gal/min	1440	gal/day
gal/min	0.06309	L/s
grad	0.90	degrees (angular)
grain	$1.4286 \ 10^{-4}$	lbm
grain/gal	142.86	lbm/MG
grain/gal	17.118	ppm
grain/gal	17.118	mg/L
hectare	2.4711	ac
hectare	10,000	m^2
hp	550.0	ft-lbf/s
hp	33,000	ft-lbf/min
hp	0.7457	kW
hp	2545	Btu/hr
hp	0.70678	Btu/s
hp-hr	2545.2	Btu
in	2.540	cm
in	0.02540	m
in	25.40	mm
in Hg	0.4910	lbf/in^2
in Hg	70.704	lbf/ft^2
in Hg	13.60	in water
in runoff	53.30	ac-ft/mi^2
in runoff	2.3230×10^6	ft^3/mi^2
in water	5.1990	lbf/ft^2
in water	0.0361	lbf/in^2
in water	0.07353	in Hg
in-lbf	0.11298	N·m
in/ft	1/0.012	mm/m
in^2	6.4516	cm^2
in^3	1/1728	ft^3
J	9.4778×10^{-4}	Btu
J	6.2415×10^{18}	eV
J	0.73756	ft-lbf

multiply	by	to obtain
J	1.0	N·m
J/s	1.0	W
kg	2.2046	lbm
kg/m³	0.06243	lbm/ft³
kip	4.4480	kN
kip	1000.0	lbf
kip	4448.0	N
kip/ft	14.594	kN/m
kip/ft²	47.880	kPa
kJ	0.94778	Btu
kJ	737.56	ft-lbf
kJ/kg	0.42992	Btu/lbm
kJ/kg·K	0.23885	Btu/lbm-°R
km	3280.8	ft
km	0.62138	mi
km/hr	0.62138	mi/hr
kN	0.2248	kips
kN·m	0.73757	ft-kips
kN/m	0.06852	kips/ft
kPa	9.8692×10^{-3}	atm
kPa	0.14504	lbf/in²
kPa	1000.0	Pa
kPa	0.02089	kips/ft²
ksi	6.8947×10^{6}	Pa
kW	737.56	ft-lbf/s
kW	44,250	ft-lbf/min
kW	1.3410	hp
kW	3413.0	Btu/hr
kW	0.9483	Btu/s
kW-hr	3413.0	Btu
kW-hr	3.60×10^{6}	J
L	1/102,790	ac-in
L	1000.0	cm³
L	0.03531	ft³
L	0.26417	gal
L	61.024	in³
L	0.0010	m³
L/s	2.1189	ft³/min
L/s	15.850	gal/min
lbf	0.001	kips
lbf	4.4482	N
lbf/ft²	0.01414	in Hg
lbf/ft²	0.19234	in water
lbf/ft²	0.00694	lbf/in²
lbf/ft²	47.880	Pa
lbf/ft²	5.0×10^{-4}	tons/ft²
lbf/in²	0.06805	atm
lbf/in²	144.0	lbf/ft²
lbf/in²	2.3080	ft water
lbf/in²	27.70	in water
lbf/in²	2.0370	in Hg
lbf/in²	6894.8	Pa
lbf/in²	0.00050	tons/in²
lbf/in²	0.0720	tons/ft²
lbm	7000.0	grains
lbm	453.59	g
lbm	0.45359	kg
lbm	4.5359×10^{5}	mg
lbm	5.0×10^{-4}	tons (mass)
lbm (of water)	0.12	gal (of water)
lbm/ac-ft-day	0.02296	lbm/1000 ft³-day
lbm/ft³	0.016018	g/cm³
lbm/ft³	16.018	kg/m³
lbm/1000 ft³-day	43.560	lbm/ac-ft-day
lbm/1000 ft³-day	133.68	lbm/MG-day
lbm/MG	0.0070	grains/gal
lbm/MG	0.11983	mg/L
lbm/MG-day	0.00748	lbm/1000 ft³-day
leagues	4428.0	m
m	1.0×10^{10}	angstroms
m	3.2808	ft
m	39.370	in
m	2.2583×10^{-4}	leagues
m	1.0936	yd
m/s	196.85	ft/min
m²	2.4711×10^{-4}	ac
m²	10.764	ft²
m²	1/10,000	hectare
m²-in	53.3	ac-ft

multiply	by	to obtain
m²-in/day	26.88	ft³/s
m³	8.1071×10^{-4}	ac-ft
m³/m·day	80.5196	gal/day-ft
m³/m²·day	24.542	gal/day-ft²
Meinzer unit	1.0	gal/day-ft²
mg	2.2046×10^{-6}	lbm
mg/L	1.0	ppm
mg/L	0.05842	grains/gal
mg/L	8.3454	lbm/MG
MG	1.0×10^{6}	gal
MG/ac-day	22.968	gal/ft²-day
MGD	1.5472	ft³/sec
MGD	1×10^{6}	gal/day
MGD/ac (mgad)	22.957	gal/day-ft²
mi	5280.0	ft
mi	80.0	chains
mi	1.6093	km
mi (statute)	0.86839	miles (nautical)
mi	320.0	rods
mi/hr	1.4667	ft/s
mi²	640.0	acres
micron	1.0×10^{-6}	m
micron	0.001	mm
mil (angular)	0.05625	degrees
mil (angular)	3.375	min
min (angular)	0.29630	mils
min (angular)	2.90888×10^{-4}	radians
min (time, mean solar)	60	s
mm	1/25.4	in
mm	1000.0	microns
mm/m	0.012	in/ft
MPa	1.0×10^{6}	Pa
N	0.22481	lbf
N	1.0×10^{5}	dynes
N·m	0.73756	ft-lbf
N·m	8.8511	in-lbf
N·m	1.0	J
N/m²	1.0	Pa
oz	28.353	g
Pa	0.001	kPa
Pa	1.4504×10^{-7}	ksi
Pa	1.4504×10^{-4}	lbf/in²
Pa	0.02089	lbf/ft²
Pa	1.0×10^{-6}	MPa
Pa	1.0	N/m²
ppm	0.05842	grains/gal
radian	180/π	degrees (angular)
radian	3437.7	min (angular)
rod	0.250	chain
rod	16.50	ft
rod	1/320	mi
s (time)	1/86,400	day (mean solar)
s (time)	1.1605×10^{-5}	day (sidereal)
s (time)	1/60	min
therm	1.0×10^{5}	Btu
ton (force)	2000.0	lbf
ton (mass)	2000.0	lbm
ton/ft²	2000.0	lbf/ft²
ton/ft²	13.889	lbf/in²
W	3.413	Btu/hr
W	0.73756	ft-lbf/s
W	1.3410×10^{-3}	hp
W	1.0	J/s
yd	1/22	chain
yd	0.91440	m
yd³	27.0	ft³
yd³	201.97	gal

Mensuration

Triangle

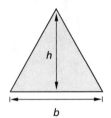 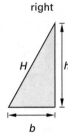 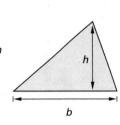

equilateral   right   oblique

$A = \frac{1}{2}bh = \frac{\sqrt{3}}{4}b^2$ $A = \frac{1}{2}bh$ $A = \frac{1}{2}bh$

$h = \frac{\sqrt{3}}{2}b$ $H^2 = b^2 + h^2$

Circle

$p = 2\pi r$

$A = \pi r^2 = \frac{p^2}{4\pi}$

Circular Segment

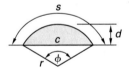

$A = \frac{1}{2}r^2(\phi - \sin\phi)$

$\phi = \frac{s}{r} = 2\left(\arccos\frac{r-d}{r}\right)$

$c = 2r\sin\left(\frac{\phi}{2}\right)$

Circular Sector

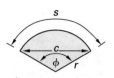

$A = \frac{1}{2}\phi r^2 = \frac{1}{2}sr$

$\phi = \frac{s}{r}$

$s = r\phi$

$c = 2r\sin\left(\frac{\phi}{2}\right)$

Parabola

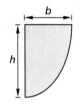

$A = \frac{2}{3}bh$

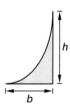

$A = \frac{1}{3}bh$

Trapezoid

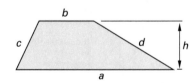

$p = a + b + c + d$

$A = \frac{1}{2}h(a + b)$

The trapezoid is isosceles if $c = d$.

Sphere

$$V = \frac{4\pi r^3}{3}$$

$$A = 4\pi r^2$$

Right Circular Cylinder

$$V = \pi r^2 h$$

$$A = 2\pi r h$$

(does not include end area)

Right Circular Cone

$$V = \frac{\pi r^2 h}{3}$$

$$A = \pi r \sqrt{r^2 + h^2}$$

(does not include base area)

PROFESSIONAL PUBLICATIONS, INC.

Background and Support

CERM Chapter 3
Algebra

> Chapter, section, equation, figure, and table numbers correspond to CERM. For additional study material, go to the corresponding chapter and section number in CERM.

9. ROOTS OF QUADRATIC EQUATIONS

$$x_1, x_2 = \frac{-b \pm \sqrt{b^2 - 4ac}}{2a} \quad [a \neq 0] \qquad 3.17$$

13. RULES FOR EXPONENTS AND RADICALS

$$(ab)^n = a^n b^n \qquad 3.24$$

$$b^{\frac{m}{n}} = \sqrt[n]{b^m} = \left(\sqrt[n]{b} \right)^m \qquad 3.25$$

$$(b^n)^m = b^{nm} \qquad 3.26$$

$$b^m b^n = b^{m+n} \qquad 3.27$$

15. LOGARITHM IDENTITIES

$$\log_b(b) = 1 \qquad 3.34$$

$$\log_b(1) = 0 \qquad 3.35$$

$$\log_b(b^n) = n \qquad 3.36$$

$$\log(x^a) = a \log(x) \qquad 3.37$$

$$\log(xy) = \log(x) + \log(y) \qquad 3.41$$

$$\log\left(\frac{x}{y}\right) = \log(x) - \log(y) \qquad 3.42$$

$$\log_a(x) = \log_b(x) \log_a(b) \qquad 3.43$$

$$\ln(x) = \log_{10}(x) \ln(10) \approx 2.3026 \; \log_{10}(x) \qquad 3.44$$

$$\log_{10}(x) = \ln(x) \log_{10}(e) \approx 0.4343 \; \ln(x) \qquad 3.45$$

CERM Chapter 4
Linear Algebra

> Chapter, section, equation, figure, and table numbers correspond to CERM. For additional study material, go to the corresponding chapter and section number in CERM.

5. DETERMINANTS

$$\mathbf{A} = \begin{bmatrix} a & b \\ c & d \end{bmatrix}$$

$$|\mathbf{A}| = \begin{vmatrix} a & b \\ c & d \end{vmatrix} = ad - bc \qquad 4.3$$

$$\mathbf{A} = \begin{bmatrix} a & b & c \\ d & e & f \\ g & h & i \end{bmatrix}$$

$$|\mathbf{A}| = a \begin{vmatrix} e & f \\ h & i \end{vmatrix} - d \begin{vmatrix} b & c \\ h & i \end{vmatrix} + g \begin{vmatrix} b & c \\ e & f \end{vmatrix} \qquad 4.6$$

CERM Chapter 5
Vectors

> Chapter, section, equation, figure, and table numbers correspond to CERM. For additional study material, go to the corresponding chapter and section number in CERM.

2. VECTORS IN *n*-SPACE

$$|\mathbf{V}| = \sqrt{(x_2 - x_1)^2 + (y_2 - y_1)^2} \qquad 5.1$$

$$\phi = \arctan\left(\frac{y_2 - y_1}{x_2 - x_1}\right) \qquad 5.2$$

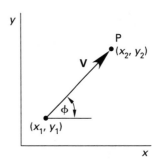

Figure 5.1 *Vector in Two-Dimensional Space*

3. UNIT VECTORS

$$\mathbf{V} = |\mathbf{V}|\,\mathbf{a} = V_x\mathbf{i} + V_y\mathbf{j} + V_z\mathbf{k} \qquad 5.8$$

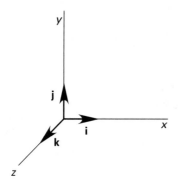

Figure 5.3 *Cartesian Unit Vectors*

8. VECTOR DOT PRODUCT

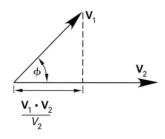

Figure 5.5 *Vector Dot Product*

$$\mathbf{V}_1 \cdot \mathbf{V}_2 = |\mathbf{V}_1|\,|\mathbf{V}_2|\cos\phi$$
$$= V_{1x}V_{2x} + V_{1y}V_{2y} + V_{1z}V_{2z} \qquad 5.21$$

9. VECTOR CROSS PRODUCT

$$|\mathbf{V}_1 \times \mathbf{V}_2| = |\mathbf{V}_1|\,|\mathbf{V}_2|\sin\phi \qquad 5.32$$

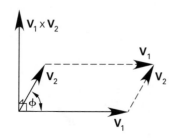

Figure 5.6 *Vector Cross Product*

CERM Chapter 6
Trigonometry

> Chapter, section, equation, figure, and table numbers correspond to CERM. For additional study material, go to the corresponding chapter and section number in CERM.

1. DEGREES AND RADIANS

multiply	by	to obtain
radians	$\dfrac{180}{\pi}$	degrees
degrees	$\dfrac{\pi}{180}$	radians

4. RIGHT TRIANGLES

$$x^2 + y^2 = r^2 \qquad 6.1$$

5. CIRCULAR TRANSCENDENTAL FUNCTIONS

$$\text{sine: } \sin\theta = \frac{y}{r} = \frac{\text{opposite}}{\text{hypotenuse}} \qquad 6.2$$

$$\text{cosine: } \cos\theta = \frac{x}{r} = \frac{\text{adjacent}}{\text{hypotenuse}} \qquad 6.3$$

$$\text{tangent: } \tan\theta = \frac{y}{x} = \frac{\text{opposite}}{\text{adjacent}} \qquad 6.4$$

$$\text{cotangent: } \cot\theta = \frac{x}{y} = \frac{\text{adjacent}}{\text{opposite}} \qquad 6.5$$

$$\text{secant: } \sec\theta = \frac{r}{x} = \frac{\text{hypotenuse}}{\text{adjacent}} \qquad 6.6$$

$$\text{cosecant: } \csc\theta = \frac{r}{y} = \frac{\text{hypotenuse}}{\text{opposite}} \qquad 6.7$$

$$\cot\theta = \frac{1}{\tan\theta} \qquad 6.8$$

$$\sec\theta = \frac{1}{\cos\theta} \qquad 6.9$$

$$\csc\theta = \frac{1}{\sin\theta} \qquad 6.10$$

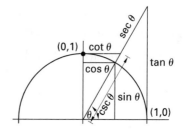

Figure 6.6 *Trigonometric Functions in a Unit Circle*

6. SMALL ANGLE APPROXIMATIONS

$$\sin\theta \approx \tan\theta \approx \theta \Big|_{\theta<10° \ (0.175 \text{ rad})} \qquad \text{6.11}$$

$$\cos\theta \approx 1 \Big|_{\theta<5° \ (0.0873 \text{ rad})} \qquad \text{6.12}$$

10. TRIGONOMETRIC IDENTITIES

$$\sin^2\theta + \cos^2\theta = 1 \qquad \text{6.14}$$

$$1 + \tan^2\theta = \sec^2\theta \qquad \text{6.15}$$

$$1 + \cot^2\theta = \csc^2\theta \qquad \text{6.16}$$

- *double-angle formulas:*

$$\sin 2\theta = 2\sin\theta\cos\theta = \frac{2\tan\theta}{1+\tan^2\theta} \qquad \text{6.17}$$

$$\cos 2\theta = \cos^2\theta - \sin^2\theta = 1 - 2\sin^2\theta$$
$$= 2\cos^2\theta - 1 = \frac{1-\tan^2\theta}{1+\tan^2\theta} \qquad \text{6.18}$$

$$\tan 2\theta = \frac{2\tan\theta}{1-\tan^2\theta} \qquad \text{6.19}$$

$$\cot 2\theta = \frac{\cot^2\theta - 1}{2\cot\theta} \qquad \text{6.20}$$

- *two-angle formulas:*

$$\sin(\theta \pm \phi) = \sin\theta\cos\phi \pm \cos\theta\sin\phi \qquad \text{6.21}$$

$$\cos(\theta \pm \phi) = \cos\theta\cos\phi \mp \sin\theta\sin\phi \qquad \text{6.22}$$

$$\tan(\theta \pm \phi) = \frac{\tan\theta \pm \tan\phi}{1 \mp \tan\theta\tan\phi} \qquad \text{6.23}$$

$$\cot(\theta \pm \phi) = \frac{\cot\phi\cot\theta \mp 1}{\cot\phi \pm \cot\theta} \qquad \text{6.24}$$

- *half-angle formulas* $(\theta < 180°)$:

$$\sin\frac{\theta}{2} = \sqrt{\frac{1-\cos\theta}{2}} \qquad \text{6.25}$$

$$\cos\frac{\theta}{2} = \sqrt{\frac{1+\cos\theta}{2}} \qquad \text{6.26}$$

$$\tan\frac{\theta}{2} = \sqrt{\frac{1-\cos\theta}{1+\cos\theta}} = \frac{\sin\theta}{1+\cos\theta} = \frac{1-\cos\theta}{\sin\theta} \qquad \text{6.27}$$

- *miscellaneous formulas* $(\theta < 90°)$:

$$\sin\theta = 2\sin\left(\frac{\theta}{2}\right)\cos\left(\frac{\theta}{2}\right) \qquad \text{6.28}$$

$$\sin\theta = \sqrt{\frac{1-\cos 2\theta}{2}} \qquad \text{6.29}$$

$$\cos\theta = \cos^2\left(\frac{\theta}{2}\right) - \sin^2\left(\frac{\theta}{2}\right) \qquad \text{6.30}$$

$$\cos\theta = \sqrt{\frac{1+\cos 2\theta}{2}} \qquad \text{6.31}$$

$$\tan\theta = \frac{2\tan\left(\frac{\theta}{2}\right)}{1-\tan^2\left(\frac{\theta}{2}\right)}$$
$$= \frac{2\left(\frac{\theta}{2}\right)\cos\left(\frac{\theta}{2}\right)}{\cos^2\left(\frac{\theta}{2}\right) - \sin^2\left(\frac{\theta}{2}\right)} \qquad \text{6.32}$$

$$\tan\theta = \sqrt{\frac{1-\cos 2\theta}{1+\cos 2\theta}}$$
$$= \frac{\sin 2\theta}{1+\cos 2\theta} = \frac{1-\cos 2\theta}{\sin 2\theta} \qquad \text{6.33}$$

$$\cot\theta = \frac{\cot^2\left(\frac{\theta}{2}\right) - 1}{2\cot\left(\frac{\theta}{2}\right)}$$
$$= \frac{\cos^2\left(\frac{\theta}{2}\right) - \sin^2\left(\frac{\theta}{2}\right)}{2\sin\left(\frac{\theta}{2}\right)\cos\left(\frac{\theta}{2}\right)} \qquad \text{6.34}$$

$$\cot\theta = \sqrt{\frac{1+\cos 2\theta}{1-\cos 2\theta}}$$
$$= \frac{1+\cos 2\theta}{\sin 2\theta} = \frac{\sin 2\theta}{1-\cos 2\theta} \qquad \text{6.35}$$

14. GENERAL TRIANGLES

$$\frac{\sin A}{a} = \frac{\sin B}{b} = \frac{\sin C}{c} \qquad \text{6.57}$$

$$a^2 = b^2 + c^2 - 2bc\cos A \qquad \text{6.58}$$

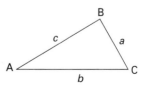

Figure 6.9 *General Triangle*

CERM Chapter 7
Analytic Geometry

> Chapter, section, equation, figure, and table numbers correspond to CERM. For additional study material, go to the corresponding chapter and section number in CERM.

10. STRAIGHT LINES

- *general form*:

$$Ax + By + C = 0 \qquad 7.8$$

$$A = -mB \qquad 7.9$$

$$B = \frac{-C}{b} \qquad 7.10$$

$$C = -aA = -bB \qquad 7.11$$

- *slope-intercept form*:

$$y = mx + b \qquad 7.12$$

$$\text{slope: } m = \frac{-A}{B} = \tan\theta = \frac{y_2 - y_1}{x_2 - x_1} \qquad 7.13$$

$$y\text{-intercept: } b = \frac{-C}{B} \qquad 7.14$$

$$x\text{-intercept: } a = \frac{-C}{A} \qquad 7.15$$

- *point-slope form*:

$$y - y_1 = m(x - x_1) \qquad 7.16$$

- *intercept form*:

$$\frac{x}{a} + \frac{y}{b} = 1 \qquad 7.17$$

- *two-point form*:

$$\frac{y - y_1}{x - x_1} = \frac{y_2 - y_1}{x_2 - x_1} \qquad 7.18$$

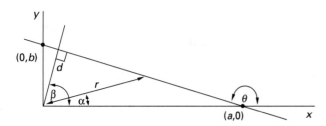

Figure 7.8 *Straight Line*

14. DISTANCES BETWEEN GEOMETRIC FIGURES

- between two points in (x, y, z) format:

$$d = \sqrt{(x_2 - x_1)^2 + (y_2 - y_1)^2 + (z_2 - z_1)^2} \qquad 7.48$$

- between a point (x_0, y_0) and a line $Ax + By + C = 0$:

$$d = \frac{|Ax_0 + By_0 + C|}{\sqrt{A^2 + B^2}} \qquad 7.49$$

- between a point (x_0, y_0, z_0) and a plane $Ax + By + Cz + D = 0$:

$$d = \frac{|Ax_0 + By_0 + Cz_0 + D|}{\sqrt{A^2 + B^2 + C^2}} \qquad 7.50$$

- between two parallel lines $Ax + By + C = 0$:

$$d = \left| \frac{|C_2|}{\sqrt{A_2^2 + B_2^2}} - \frac{|C_1|}{\sqrt{A_1^2 + B_1^2}} \right| \qquad 7.51$$

17. CIRCLE

$$Ax^2 + Ay^2 + Dx + Ey + F = 0 \qquad 7.66$$

$$(x - h)^2 + (y - k)^2 = r^2 \quad \text{[center at (h,k)]} \qquad 7.67$$

18. PARABOLA

$$(y - k)^2 = 4p(x - h) \Big|_{\text{opens horizontally}} \qquad 7.73$$
$$\text{[vertex at (h,k)]}$$

$$(x - h)^2 = 4p(y - k) \Big|_{\text{opens vertically}} \qquad 7.75$$
$$\text{[vertex at (h,k)]}$$

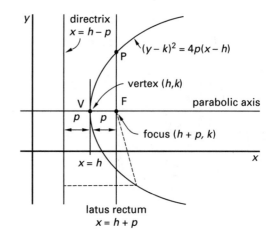

Figure 7.13 *Parabola*

CERM Chapter 11
Probability and Statistical Analysis of Data

> Chapter, section, equation, figure, and table numbers correspond to CERM. For additional study material, go to the corresponding chapter and section number in CERM.

12. POISSON DISTRIBUTION

$$p\{x\} = f(x) = \frac{e^{-\lambda}\lambda^x}{x!} \qquad [\lambda > 0] \qquad 11.34$$

λ is both the mean and the variance of the Poisson distribution.

15. NORMAL DISTRIBUTION

$$z = \frac{x_0 - \mu}{\sigma} \qquad \text{[standard normal variable]} \qquad 11.43$$

16. APPLICATION: RELIABILITY

$$R\{t\} = e^{-\lambda t} = e^{\frac{-t}{\text{MTTF}}} \qquad 11.47$$

$$\lambda = \frac{1}{\text{MTTF}} \qquad 11.48$$

$$R\{t\} = 1 - F(t) = 1 - (1 - e^{-\lambda t}) = e^{-\lambda t} \qquad 11.49$$

$$z\{t\} = \lambda \qquad \text{[hazard function]} \qquad 11.50$$

18. MEASURES OF CENTRAL TENDENCY

$$\bar{x} = \left(\frac{1}{n}\right)(x_1 + x_2 + \cdots + x_n) = \frac{\sum x_i}{n} \qquad 11.56$$

$$\text{geometric mean} = \sqrt[n]{x_1 x_2 x_3 \cdots x_n} \qquad [x_i > 0] \quad 11.57$$

$$\text{harmonic mean} = \frac{n}{\dfrac{1}{x_1} + \dfrac{1}{x_2} + \cdots + \dfrac{1}{x_n}} \qquad 11.58$$

$$\text{root mean square} = x_{\text{rms}} = \sqrt{\frac{\sum x_i^2}{n}} \qquad 11.59$$

19. MEASURES OF DISPERSION

$$\sigma = \sqrt{\frac{\sum(x_i - \mu)^2}{N}} = \sqrt{\frac{\sum x_i^2}{N} - \mu^2} \qquad 11.60$$

$$s = \sqrt{\frac{\sum(x_i - \bar{x})^2}{n-1}} = \sqrt{\frac{\sum x_i^2 - \dfrac{(\sum x_i)^2}{n}}{n-1}} \qquad 11.61$$

$$\sigma_{\text{sample}} = s\sqrt{\frac{n-1}{n}} \qquad 11.62$$

$$\text{coefficient of variation} = \frac{s}{\bar{x}} \qquad 11.63$$

22. APPLICATION: CONFIDENCE LIMITS

$$\text{LCL: } \mu - z_c\sigma$$

$$\text{UCL: } \mu + z_c\sigma$$

Table 11.1 *Values of z for Various Confidence Levels*

confidence level, C	one-tail limit z	two-tail limit z
90%	1.28	1.645
95%	1.645	1.96
97.5%	1.96	2.17
99%	2.33	2.575
99.5%	2.575	2.81
99.75%	2.81	3.00

Areas Under the Standard Normal Curve

$(0 \text{ to } z)$

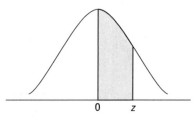

z	0	1	2	3	4	5	6	7	8	9
0.0	0.0000	0.0040	0.0080	0.0120	0.0160	0.0199	0.0239	0.0279	0.0319	0.0359
0.1	0.0398	0.0438	0.0478	0.0517	0.0557	0.0596	0.0636	0.0675	0.0714	0.0754
0.2	0.0793	0.0832	0.0871	0.0910	0.0948	0.0987	0.1026	0.1064	0.1103	0.1141
0.3	0.1179	0.1217	0.1255	0.1293	0.1331	0.1368	0.1406	0.1443	0.1480	0.1517
0.4	0.1554	0.1591	0.1628	0.1664	0.1700	0.1736	0.1772	0.1808	0.1844	0.1879
0.5	0.1915	0.1950	0.1985	0.2019	0.2054	0.2088	0.2123	0.2157	0.2190	0.2224
0.6	0.2258	0.2291	0.2324	0.2357	0.2389	0.2422	0.2454	0.2486	0.2518	0.2549
0.7	0.2580	0.2612	0.2642	0.2673	0.2704	0.2734	0.2764	0.2794	0.2823	0.2852
0.8	0.2881	0.2910	0.2939	0.2967	0.2996	0.3023	0.3051	0.3078	0.3106	0.3133
0.9	0.3159	0.3186	0.3212	0.3238	0.3264	0.3289	0.3315	0.3340	0.3365	0.3389
1.0	0.3413	0.3438	0.3461	0.3485	0.3508	0.3531	0.3554	0.3577	0.3599	0.3621
1.1	0.3643	0.3665	0.3686	0.3708	0.3729	0.3749	0.3770	0.3790	0.3810	0.3830
1.2	0.3849	0.3869	0.3888	0.3907	0.3925	0.3944	0.3962	0.3980	0.3997	0.4015
1.3	0.4032	0.4049	0.4066	0.4082	0.4099	0.4115	0.4131	0.4147	0.4162	0.4177
1.4	0.4192	0.4207	0.4222	0.4236	0.4251	0.4265	0.4279	0.4292	0.4306	0.4319
1.5	0.4332	0.4345	0.4357	0.4370	0.4382	0.4394	0.4406	0.4418	0.4429	0.4441
1.6	0.4452	0.4463	0.4474	0.4484	0.4495	0.4505	0.4515	0.4525	0.4535	0.4545
1.7	0.4554	0.4564	0.4573	0.4582	0.4591	0.4599	0.4608	0.4616	0.4625	0.4633
1.8	0.4641	0.4649	0.4656	0.4664	0.4671	0.4678	0.4686	0.4693	0.4699	0.4706
1.9	0.4713	0.4719	0.4726	0.4732	0.4738	0.4744	0.4750	0.4756	0.4761	0.4767
2.0	0.4772	0.4778	0.4783	0.4788	0.4793	0.4798	0.4803	0.4808	0.4812	0.4817
2.1	0.4821	0.4826	0.4830	0.4834	0.4838	0.4842	0.4846	0.4850	0.4854	0.4857
2.2	0.4861	0.4864	0.4868	0.4871	0.4875	0.4878	0.4881	0.4884	0.4887	0.4890
2.3	0.4893	0.4896	0.4898	0.4901	0.4904	0.4906	0.4909	0.4911	0.4913	0.4916
2.4	0.4918	0.4920	0.4922	0.4925	0.4927	0.4929	0.4931	0.4932	0.4934	0.4936
2.5	0.4938	0.4940	0.4941	0.4943	0.4945	0.4946	0.4948	0.4949	0.4951	0.4952
2.6	0.4953	0.4955	0.4956	0.4957	0.4959	0.4960	0.4961	0.4962	0.4963	0.4964
2.7	0.4965	0.4966	0.4967	0.4968	0.4969	0.4970	0.4971	0.4972	0.4973	0.4974
2.8	0.4974	0.4975	0.4976	0.4977	0.4977	0.4978	0.4979	0.4979	0.4980	0.4981
2.9	0.4981	0.4982	0.4982	0.4983	0.4984	0.4984	0.4985	0.4985	0.4986	0.4986
3.0	0.4987	0.4987	0.4987	0.4988	0.4988	0.4989	0.4989	0.4989	0.4990	0.4990
3.1	0.4990	0.4991	0.4991	0.4991	0.4992	0.4992	0.4992	0.4992	0.4993	0.4993
3.2	0.4993	0.4993	0.4994	0.4994	0.4994	0.4994	0.4994	0.4995	0.4995	0.4995
3.3	0.4995	0.4995	0.4996	0.4996	0.4996	0.4996	0.4996	0.4996	0.4996	0.4997
3.4	0.4997	0.4997	0.4997	0.4997	0.4997	0.4997	0.4997	0.4997	0.4997	0.4998
3.5	0.4998	0.4998	0.4998	0.4998	0.4998	0.4998	0.4998	0.4998	0.4998	0.4998
3.6	0.4998	0.4998	0.4999	0.4999	0.4999	0.4999	0.4999	0.4999	0.4999	0.4999
3.7	0.4999	0.4999	0.4999	0.4999	0.4999	0.4999	0.4999	0.4999	0.4999	0.4999
3.8	0.4999	0.4999	0.4999	0.4999	0.4999	0.4999	0.4999	0.4999	0.4999	0.4999
3.9	0.5000	0.5000	0.5000	0.5000	0.5000	0.5000	0.5000	0.5000	0.5000	0.5000

CERM Chapter 13
Energy, Work, and Power

Chapter, section, equation, figure, and table numbers correspond to CERM. For additional study material, go to the corresponding chapter and section number in CERM.

3. WORK

$$W_{\text{variable force}} = \int \mathbf{F} \cdot d\mathbf{s} \quad \text{[linear systems]} \qquad 13.4$$

$$W_{\text{variable torque}} = \int \mathbf{T} \cdot d\boldsymbol{\theta} \quad \text{[rotational systems]} \qquad 13.5$$

$$W_{\text{constant force}} = \mathbf{F} \cdot \mathbf{s} = Fs \, \cos\phi \quad \text{[linear systems]} \quad 13.6$$

$$W_{\text{constant torque}} = \mathbf{T} \cdot \boldsymbol{\theta} = Fr\theta \, \cos\phi$$
$$\text{[rotational systems]} \quad 13.7$$

$$W_{\text{friction}} = F_f s \qquad 13.8$$

$$W_{\text{gravity}} = \left(\frac{mg}{g_c}\right)(h_2 - h_1) \qquad 13.9(b)$$

$$W_{\text{spring}} = \tfrac{1}{2}k\left(\delta_2^2 - \delta_1^2\right) \qquad 13.10$$

4. POTENTIAL ENERGY OF A MASS

$$E_{\text{potential}} = \frac{mgh}{g_c} \qquad 13.11(b)$$

5. KINETIC ENERGY OF A MASS

$$E_{\text{kinetic}} = \frac{m\text{v}^2}{2g_c} \qquad 13.13(b)$$

$$E_{\text{rotational}} = \frac{I\omega^2}{2g_c} \qquad 13.15(b)$$

6. SPRING ENERGY

$$E_{\text{spring}} = \tfrac{1}{2}k\delta^2 \qquad 13.16$$

8. INTERNAL ENERGY OF A MASS

$$Q = mc\,\Delta T \qquad 13.19$$

9. WORK-ENERGY PRINCIPLE

$$W = \Delta E = E_2 - E_1 \qquad 13.24$$

11. POWER

$$P = \frac{W}{\Delta t} \qquad 13.25$$

$$P = F\text{v} \quad \text{[linear systems]} \qquad 13.26$$

$$P = T\omega \quad \text{[rotational systems]} \qquad 13.27$$

Water Resources

CERM Chapter 14
Fluid Properties

Chapter, section, equation, figure, and table numbers correspond to CERM. For additional study material, go to the corresponding chapter and section number in CERM.

4. DENSITY

$$\rho = \frac{p}{RT} \quad \text{[gases]} \qquad 14.5$$

5. SPECIFIC VOLUME

$$v = \frac{1}{\rho} \qquad 14.6$$

6. SPECIFIC GRAVITY

$$SG_{liquid} = \frac{\rho_{liquid}}{\rho_{water}} \qquad 14.7$$

$$SG_{gas} = \frac{\rho_{gas}}{\rho_{air}} \qquad 14.8$$

$$SG_{gas} = \frac{MW_{gas}}{MW_{air}} = \frac{MW_{gas}}{29.0}$$

$$= \frac{R_{air}}{R_{gas}} = \frac{53.3 \, \frac{\text{ft-lbf}}{\text{lbm-}°R}}{R_{gas}} \qquad 14.9$$

7. SPECIFIC WEIGHT

$$\gamma = \rho \times \frac{g}{g_c} \qquad 14.14(b)$$

9. VISCOSITY

$$\tau = \mu \frac{dv}{dy} \qquad 14.18$$

The constant of proportionality, μ, is the absolute viscosity.

10. KINEMATIC VISCOSITY

$$\nu = \frac{\mu g_c}{\rho} \qquad 14.19(b)$$

18. BULK MODULUS

The bulk modulus, E, is the reciprocal of compressibility, β.

$$E = \frac{1}{\beta} \qquad 14.36$$

19. SPEED OF SOUND

$$a = \sqrt{\frac{Eg_c}{\rho}} = \sqrt{\frac{g_c}{\beta\rho}} \quad \text{[liquids]} \qquad 14.37(b)$$

$$a = \sqrt{\frac{Eg_c}{\rho}} = \sqrt{\frac{kg_c p}{\rho}}$$

$$= \sqrt{kg_c RT} = \sqrt{\frac{kg_c R^* T}{MW}} \quad \text{[gases]} \qquad 14.38(b)$$

CERM Chapter 15
Fluid Statics

Chapter, section, equation, figure, and table numbers correspond to CERM. For additional study material, go to the corresponding chapter and section number in CERM.

2. MANOMETERS

$$p_2 - p_1 = \rho_m \times \frac{g}{g_c} \times h$$
$$= \gamma_m h \qquad 15.4$$
$$p_2 - p_1 = \frac{g}{g_c} \times (\rho_m h + \rho_1 h_1 - \rho_2 h_2)$$
$$= \gamma_m h + \gamma_1 h_1 - \gamma_2 h_2 \qquad 15.5(b)$$

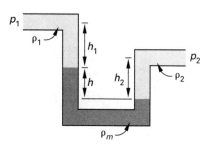

Figure 15.4 *Manometer Requiring Corrections*

3. HYDROSTATIC PRESSURE

$$p = \frac{F}{A} \qquad\qquad 15.6$$

4. FLUID HEIGHT EQUIVALENT TO PRESSURE

$$p = \frac{\rho g h}{g_c} = \gamma h \qquad\qquad 15.7(b)$$

7. PRESSURE ON A VERTICAL PLANE SURFACE

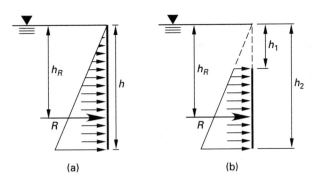

Figure 15.9 *Hydrostatic Pressure on a Vertical Plane Surface*

$$\bar{p} = \frac{\frac{1}{2}\rho g(h_1 + h_2)}{g_c} = \frac{1}{2}\gamma(h_1 + h_2) \qquad 15.15(b)$$

$$R = \bar{p}A \qquad\qquad 15.16$$

$$h_R = \frac{2}{3}\left(h_1 + h_2 - \frac{h_1 h_2}{h_1 + h_2}\right) \qquad 15.17$$

9. PRESSURE ON A GENERAL PLANE SURFACE

$$\bar{p} = \frac{\rho g h_c \sin\theta}{g_c} = \gamma h_c \sin\theta \qquad 15.24(b)$$

$$R = \bar{p}A \qquad\qquad 15.25$$

$$h_R = h_c + \frac{I_c}{Ah_c} \qquad\qquad 15.26$$

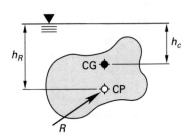

Figure 15.11 *General Plane Surface*

13. HYDROSTATIC FORCES ON A DAM

$$M_{\text{overturning}} = R_x \times y_{R_x} \qquad 15.31$$

$$M_{\text{resisting}} = (R_y \times x_{R_y}) + (W \times x_{\text{CG}}) \qquad 15.32$$

$$(\text{FS})_{\text{overturning}} = \frac{M_{\text{resisting}}}{M_{\text{overturning}}} \qquad 15.33$$

$$\begin{aligned} F_f &= \mu_{\text{static}}\ N \\ &= \mu_{\text{static}}(W + R_y) \end{aligned} \qquad 15.34$$

$$(\text{FS})_{\text{sliding}} = \frac{F_f}{R_x} \qquad 15.35$$

$$p_{\text{max}}, p_{\text{min}} = \left(\frac{R_y + W}{b}\right)\left(1 \pm \frac{6e}{b}\right)\left[\begin{smallmatrix}\text{per unit}\\\text{width}\end{smallmatrix}\right] \quad 15.36$$

$$e = \frac{b}{2} - x_{\text{v}} \qquad 15.37$$

$$x_{\text{v}} = \frac{M_{\text{resisting}}}{R_y + W} \qquad 15.38$$

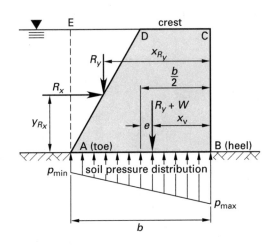

Figure 15.16 *Dam*

18. BUOYANCY

$$F_{\text{buoyant}} = \frac{\rho g V_{\text{displaced}}}{g_c} = \gamma V_{\text{displaced}} \qquad 15.56(b)$$

$$\text{SG} = \frac{W_{\text{dry}}}{W_{\text{dry}} - W_{\text{submerged}}} \qquad 15.57$$

CERM Chapter 16
Fluid Flow Parameters

Chapter, section, equation, figure, and table numbers correspond to CERM. For additional study material, go to the corresponding chapter and section number in CERM.

2. KINETIC ENERGY

$$E_v = \frac{v^2}{2g_c} \qquad \text{16.3(b)}$$

3. POTENTIAL ENERGY

$$E_z = \frac{zg}{g_c} \qquad \text{16.5(b)}$$

4. PRESSURE ENERGY

$$E_p = \frac{p}{\rho} \qquad \text{16.6}$$

5. BERNOULLI EQUATION

$$E_t = \frac{p}{\rho} + \frac{v^2}{2g_c} + \frac{zg}{g_c} \qquad \text{16.11(b)}$$

8. HYDRAULIC RADIUS

$$r_h = \frac{\text{area in flow}}{\text{wetted perimeter}} = \frac{A}{s} \qquad \text{16.18}$$

9. EQUIVALENT DIAMETER

$$D_e = 4r_h \qquad \text{16.20}$$

Table 16.1 *Equivalent Diameters for Common Conduit Shapes*

conduit cross section	D_e
flowing full	
annulus (outer diameter D_o, inner diameter D_i)	$D_o - D_i$
square (side L)	L
rectangle (sides L_1 and L_2)	$\dfrac{2L_1 L_2}{L_1 + L_2}$
flowing partially full	
half-filled circle (diameter D)	D
rectangle (h deep, L wide)	$\dfrac{4hL}{L + 2h}$
wide, shallow stream (h deep)	$4h$
triangle (h deep, L broad, s side)	$\dfrac{hL}{s}$
trapezoid (h deep, a wide at top, b wide at bottom, s side)	$\dfrac{2h(a + b)}{b + 2s}$

10. REYNOLDS NUMBER

$$\text{Re} = \frac{D_e v \rho}{g_c \mu} \qquad \text{16.22(b)}$$

$$\text{Re} = \frac{D_e v}{\nu} \qquad \text{16.23}$$

16. SPECIFIC ENERGY

$$E_{\text{specific}} = E_p + E_v \qquad \text{16.36}$$

$$E_{\text{specific}} = \frac{p}{\rho} + \frac{v^2}{2g_c} \qquad \text{16.37(b)}$$

CERM Chapter 17
Fluid Dynamics

Chapter, section, equation, figure, and table numbers correspond to CERM. For additional study material, go to the corresponding chapter and section number in CERM.

9. ENERGY LOSS DUE TO FRICTION: TURBULENT FLOW

- *Darcy equation*

$$h_f = \frac{fLv^2}{2Dg} \qquad \text{17.28}$$

- *Hazen-Williams equation*

$$h_{f,\text{feet}} = \frac{3.022 v^{1.85} L}{C^{1.85} D^{1.17}} \qquad 17.30$$

$$h_{f,\text{feet}} = \frac{10.44 L \dot{V}_{\text{gpm}}^{1.85}}{C^{1.85} d_{\text{inches}}^{4.87}} \qquad 17.31$$

15. MINOR LOSSES

$$h_m = K h_v \qquad 17.41$$

$$K = \frac{f L_e}{D} \qquad 17.42$$

- *sudden enlargements:* (D_1 is the smaller of the two diameters)

$$K = \left(1 - \left(\frac{D_1}{D_2}\right)^2\right)^2 \qquad 17.43$$

- *sudden contractions:* (D_1 is the smaller of the two diameters)

$$K = \frac{1}{2}\left(1 - \left(\frac{D_1}{D_2}\right)^2\right) \qquad 17.44$$

16. VALVE FLOW COEFFICIENTS

$$Q_{\text{gpm}} = C_v \sqrt{\frac{\Delta p_{\text{psi}}}{\text{SG}}} \qquad 17.50$$

21. DISCHARGE FROM TANKS

$$v_t = \sqrt{2gh} \qquad 17.67$$

$$h = z_1 - z_2 \qquad 17.68$$

$$v_o = C_v \sqrt{2gh} \qquad 17.69$$

$$C_v = \frac{\text{actual velocity}}{\text{theoretical velocity}} = \frac{v_o}{v_t} \qquad 17.70$$

$$C_c = \frac{\text{area of vena contracta}}{\text{orifice area}} \qquad 17.74$$

$$C_d = C_v C_c = \frac{\text{actual discharge}}{\text{theoretical discharge}} \qquad 17.76$$

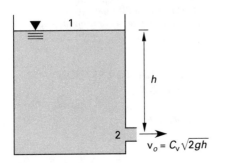

Figure 17.10 *Discharge from a Tank*

23. COORDINATES OF A FLUID STREAM

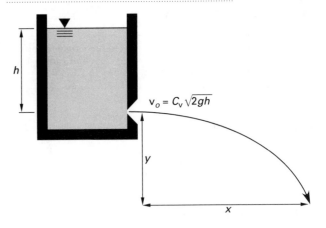

Figure 17.13 *Coordinates of a Fluid Stream*

$$v_x = v_o \qquad 17.78$$

$$x = v_o t = v_o \sqrt{\frac{2y}{g}} = 2 C_v \sqrt{hy} \qquad 17.79$$

$$v_y = gt \qquad 17.80$$

$$y = \frac{gt^2}{2} = \frac{gx^2}{2v_o^2} = \frac{x^2}{4hC_v^2} \qquad 17.81$$

33. PITOT-STATIC GAUGE

$$\frac{v^2}{2g_c} = \frac{p_t - p_s}{\rho} = \frac{h(\rho_m - \rho)}{\rho} \times \left(\frac{g}{g_c}\right) \qquad 17.144(b)$$

$$v = \sqrt{\frac{2gh(\rho_m - \rho)}{\rho}} \qquad 17.145$$

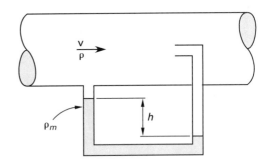

Figure 17.23 *Pitot-Static Gauge*

34. VENTURI METER

$$v_2 = \left(\frac{C_v}{\sqrt{1 - \left(\frac{A_2}{A_1}\right)^2}} \right) \sqrt{\frac{2g_c(p_1 - p_2)}{\rho}}$$ 17.148(b)

$$\beta = \frac{D_2}{D_1}$$ 17.149

$$F_{va} = \frac{1}{\sqrt{1 - \left(\frac{A_2}{A_1}\right)^2}} = \frac{1}{\sqrt{1 - \beta^4}}$$ 17.150

$$v_2 = \left(\frac{C_v}{\sqrt{1 - \beta^4}} \right) \sqrt{\frac{2g(\rho_m - \rho)h}{\rho}}$$

$$= C_v F_{va} \sqrt{\frac{2g(\rho_m - \rho)h}{\rho}}$$ 17.151

$$\dot{V} = C_d A_2 v_{2,\text{ideal}}$$ 17.152

$$C_f = C_d F_{va} = \frac{C_d}{\sqrt{1 - \beta^4}}$$ 17.153

$$\dot{V} = C_f A_2 \sqrt{\frac{2g(\rho_m - \rho)h}{\rho}}$$ 17.154

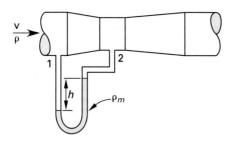

Figure 17.25 *Venturi Meter with Manometer*

35. ORIFICE METER

$$A_2 = C_c A_o$$ 17.155

$$v_o = \left(\frac{C_v}{\sqrt{1 - \left(\frac{C_c A_o}{A_1}\right)^2}} \right) \sqrt{\frac{2g_c(p_1 - p_2)}{\rho}}$$ 17.156(b)

$$v_o = \left(\frac{C_v}{\sqrt{1 - \left(\frac{C_c A_o}{A_1}\right)^2}} \right) \sqrt{\frac{2g(\rho_m - \rho)h}{\rho}}$$ 17.157

$$F_{va} = \frac{1}{\sqrt{1 - \left(\frac{C_c A_o}{A_1}\right)^2}}$$ 17.158

$$C_f = C_d F_{va}$$ 17.159

$$\dot{V} = C_f A_o \sqrt{\frac{2g(\rho_m - \rho)h}{\rho}}$$

$$= C_f A_o \sqrt{\frac{2g(p_1 - p_2)}{\rho}}$$ 17.160

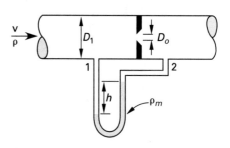

Figure 17.27 *Orifice Meter with Differential Manometer*

47. CONFINED STREAMS IN PIPE BENDS

$$F_x = p_2 A_2 \, \cos\theta - p_1 A_1 \\ + \frac{\dot{m}(v_2 \, \cos\theta - v_1)}{g_c}$$ 17.200(b)

$$F_y = \left(p_2 A_2 + \frac{\dot{m} v_2}{g_c} \right) \sin\theta$$ 17.201(b)

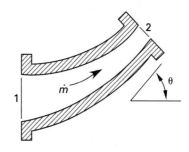

Figure 17.40 *Pipe Bend*

48. WATER HAMMER

$$t = \frac{L}{a} \quad \text{[one way]}$$ 17.202

$$t = \frac{2L}{a} \quad \text{[round trip]}$$ 17.203

$$\Delta p = \frac{\rho a \Delta v}{g_c}$$ 17.204(b)

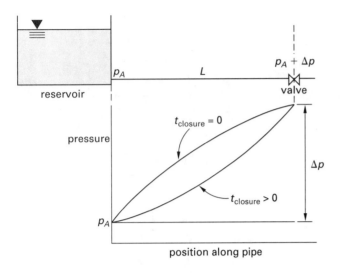

Figure 17.41 *Water Hammer*

52. DRAG

$$F_D = \frac{C_D A \rho v^2}{2g_c} \qquad \text{17.212(b)}$$

54. TERMINAL VELOCITY

$$v = \sqrt{\frac{2(mg - F_b)}{C_D A \rho_{\text{fluid}}}}$$

$$= \sqrt{\frac{2Vg(\rho_{\text{object}} - \rho_{\text{fluid}})}{C_D A \, \rho_{\text{fluid}}}} \qquad \text{17.216}$$

$$v = \sqrt{\frac{4Dg(\rho_{\text{sphere}} - \rho_{\text{fluid}})}{3C_D \rho_{\text{fluid}}}} \quad [\text{sphere}] \qquad \text{17.217}$$

$$v = \frac{D^2(\rho_{\text{sphere}} - \rho_{\text{fluid}})}{18\,\mu} \times \left(\frac{g}{g_c}\right)$$

$$\begin{bmatrix} \text{small particles;} \\ \text{Stokes' law} \end{bmatrix} \qquad \text{17.218(b)}$$

58. SIMILARITY

$$L_r = \frac{\text{size of model}}{\text{size of prototype}} \qquad \text{17.223}$$

59. VISCOUS AND INERTIAL FORCES DOMINATE

$$\text{Re}_m = \text{Re}_p \qquad \text{17.226}$$

$$\frac{L_m v_m}{\nu_m} = \frac{L_p v_p}{\nu_p} \qquad \text{17.227}$$

60. INERTIAL AND GRAVITATIONAL FORCES DOMINATE

$$\text{Fr} = \frac{v^2}{Lg} \qquad \text{17.228}$$

$$\text{Re}_m = \text{Re}_p \qquad \text{17.229}$$

$$\text{Fr}_m = \text{Fr}_p \qquad \text{17.230}$$

$$\frac{\nu_m}{\nu_p} = \left(\frac{L_m}{L_p}\right)^{\frac{3}{2}} = (L_r)^{\frac{3}{2}} \qquad \text{17.231}$$

$$n_r = (L_r)^{\frac{1}{6}} \qquad \text{17.232}$$

61. SURFACE TENSION FORCE DOMINATES

$$\text{We} = \frac{v^2 L \rho}{\sigma} \qquad \text{17.233}$$

$$\text{We}_m = \text{We}_p \qquad \text{17.234}$$

CERM Chapter 18
Hydraulic Machines

Chapter, section, equation, figure, and table numbers correspond to CERM. For additional study material, go to the corresponding chapter and section number in CERM.

6. TYPES OF CENTRIFUGAL PUMPS

$$v_{\text{tip}} = \frac{\pi D n}{60 \, \frac{\text{sec}}{\text{min}}} = \frac{D\omega}{2} \qquad \text{18.3}$$

10. PUMPING POWER

Table 18.5 *Hydraulic Horsepower Equations[a]*

	Q (gal/min)	$\dot{m}$ (lbm/sec)	$\dot{V}$ (ft^3/sec)
h_A in feet	$\dfrac{h_A Q(\text{SG})}{3956}$	$\left(\dfrac{h_A \dot{m}}{550}\right) \times \left(\dfrac{g}{g_c}\right)$	$\dfrac{h_A \dot{V}(\text{SG})}{8.814}$
Δp in psi[b]	$\dfrac{\Delta p Q}{1714}$	$\left[\dfrac{\Delta p \dot{m}}{(238.3)(\text{SG})}\right] \times \left(\dfrac{g}{g_c}\right)$	$\dfrac{\Delta p \dot{V}}{3.819}$
Δp in psf[b]	$\dfrac{\Delta p Q}{2.468 \times 10^5}$	$\left[\dfrac{\Delta p \dot{m}}{(34{,}320)(\text{SG})}\right] \times \left(\dfrac{g}{g_c}\right)$	$\dfrac{\Delta p \dot{V}}{550}$
W in $\dfrac{\text{ft-lbf}}{\text{lbm}}$	$\dfrac{W Q(\text{SG})}{3956}$	$\dfrac{W \dot{m}}{550}$	$\dfrac{W \dot{V}(\text{SG})}{8.814}$

(Multiply horsepower by 0.7457 to obtain kilowatts.)
[a]Table 18.5 is based on $\rho_{\text{water}} = 62.4$ lbm/ft^3 and $g = 32.2$ ft/sec^2.
[b]Velocity head changes must be included in Δp.

Table 18.6 Hydraulic Kilowatt Equations[a]

	Q (L/s)	$\dot{m}$ (kg/s)	$\dot{V}$ (m³/s)
h_A in meters	$\dfrac{(9.81)h_A\,Q(\text{SG})}{1000}$	$\dfrac{(9.81)h_A\,\dot{m}}{1000}$	$(9.81)h_A\,\dot{V}(\text{SG})$
Δp in kPa[b]	$\dfrac{\Delta p\,Q}{1000}$	$\dfrac{\Delta p\,\dot{m}}{1000(\text{SG})}$	$\Delta p\,\dot{V}$
W in $\dfrac{\text{J}}{\text{kg}}$[b]	$\dfrac{W Q(\text{SG})}{1000}$	$\dfrac{W \dot{m}}{1000}$	$W\dot{V}(\text{SG})$

(Multiply kilowatts by 1.341 to obtain horsepower.)
[a]Table 18.6 is based on $\rho_{\text{water}} = 1000$ kg/m³ and $g = 9.81$ m/s².
[b]Velocity head changes must be included in Δp.

11. PUMPING EFFICIENCY

$$\eta = \eta_p \eta_m = \frac{\text{hydraulic horsepower}}{\text{motor horsepower}} \qquad 18.16$$

12. COST OF ELECTRICITY

$$\text{cost} = \frac{W_{\text{kW-hr}} \times \text{cost per kW-hr}}{\eta_m} \qquad 18.18$$

13. STANDARD MOTOR SIZES AND SPEEDS

$$n = \frac{120 \times f}{\text{no. of poles}} \quad [\text{synchronous speed}] \qquad 18.19$$

$$\begin{array}{l}\text{slip} \\ \text{(in rpm)}\end{array} = \text{synchronous speed} - \text{actual speed} \quad 18.20$$

$$\begin{array}{l}\text{percent} \\ \text{slip}\end{array} = 100\% \times \frac{\begin{array}{c}\text{synchronous speed} \\ -\text{ actual speed}\end{array}}{\text{synchronous speed}} \qquad 18.21$$

$$\text{kVA rating} = \frac{\text{motor power in kW}}{\text{power factor}} \qquad 18.22$$

14. PUMP SHAFT LOADING

$$T_{\text{in-lbf}} = \frac{63{,}025 \times P_{\text{hp}}}{n} \qquad 18.23$$

$$T_{\text{ft-lbf}} = \frac{5252 \times P_{\text{hp}}}{n} \qquad 18.24$$

$$\text{overhung load} = \frac{2KT}{D_{\text{sheave}}} \qquad 18.27$$

15. SPECIFIC SPEED

$$n_s = \frac{n\sqrt{Q}}{(h_A)^{0.75}} \qquad 18.28(b)$$

17. NET POSITIVE SUCTION HEAD (NPSH)

$$\frac{\text{NPSHR}_2}{\text{NPSHR}_1} = \left(\frac{Q_2}{Q_1}\right)^2 \qquad 18.29$$

$$\text{NPSHA} = h_{\text{atm}} + h_{z(s)} - h_{f(s)} - h_{\text{vp}} - h_{\text{ac}} \qquad 18.31$$

$$\text{NPSHA} = h_{p(s)} + h_{v(s)} - h_{\text{vp}} - h_{\text{ac}} \qquad 18.32$$

$$\text{NPSHA} < \text{NPSHR} \quad [\text{criterion for cavitation}] \qquad 18.33$$

20. SUCTION SPECIFIC SPEED

$$n_{\text{ss}} = \frac{n\sqrt{Q}}{(\text{NPSHR in ft})^{0.75}} \qquad 18.36(b)$$

22. SYSTEM CURVES

$$\frac{h_{f,1}}{h_{f,2}} = \left(\frac{Q_1}{Q_2}\right)^2 \qquad 18.40$$

26. AFFINITY LAWS

(Impeller size is constant; speed is varied.)

$$\frac{Q_2}{Q_1} = \frac{n_2}{n_1} \qquad 18.41$$

$$\frac{h_2}{h_1} = \left(\frac{n_2}{n_1}\right)^2 = \left(\frac{Q_2}{Q_1}\right)^2 \qquad 18.42$$

$$\frac{P_2}{P_1} = \left(\frac{n_2}{n_1}\right)^3 = \left(\frac{Q_2}{Q_1}\right)^3 \qquad 18.43$$

$$\frac{Q_2}{Q_1} = \frac{D_2}{D_1} \qquad 18.44$$

$$\frac{h_2}{h_1} = \left(\frac{D_2}{D_1}\right)^2 \qquad 18.45$$

$$\frac{P_2}{P_1} = \left(\frac{D_2}{D_1}\right)^3 \qquad 18.46$$

27. PUMP SIMILARITY

(Impeller size varies.)

$$\frac{n_1 D_1}{\sqrt{h_1}} = \frac{n_2 D_2}{\sqrt{h_2}} \qquad 18.48$$

$$\frac{Q_1}{D_1^2 \sqrt{h_1}} = \frac{Q_2}{D_2^2 \sqrt{h_2}} \qquad 18.49$$

$$\frac{P_1}{\rho_1 D_1^2 h_1^{1.5}} = \frac{P_2}{\rho_2 D_2^2 h_2^{1.5}} \qquad 18.50$$

$$\frac{Q_1}{n_1 D_1^3} = \frac{Q_2}{n_2 D_2^3} \qquad 18.51$$

$$\frac{P_1}{\rho_1 n_1^3 D_1^5} = \frac{P_2}{\rho_2 n_2^3 D_2^5} \qquad \text{18.52}$$

$$\frac{n_1 \sqrt{Q_1}}{(h_1)^{0.75}} = \frac{n_2 \sqrt{Q_2}}{(h_2)^{0.75}} \qquad \text{18.53}$$

CERM Chapter 19
Open Channel Flow

Chapter, section, equation, figure, and table numbers correspond to CERM. For additional study material, go to the corresponding chapter and section number in CERM.

5. PARAMETERS USED IN OPEN CHANNEL FLOW

- *hydraulic radius*

$$R = \frac{A}{P} \qquad \text{19.2}$$

- *hydraulic depth*

$$D_h = \frac{A}{w} \qquad \text{19.3}$$

- *energy gradient* (geometric slope)

$$S_0 = \frac{\Delta z}{L} \quad \text{[uniform flow]} \qquad \text{19.7}$$

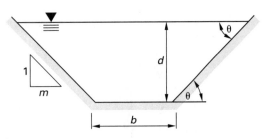

Figure 19.2 *Trapezoidal Cross Section*

6. GOVERNING EQUATIONS FOR UNIFORM FLOW

$n =$ Manning constant

$$C = \left(\frac{1.49}{n}\right) R^{\frac{1}{6}} \quad \text{[Chezy equation]} \qquad \text{19.11(b)}$$

$$v = \left(\frac{1.49}{n}\right) R^{\frac{2}{3}} \sqrt{S} \quad \left[\begin{array}{c}\text{Chezy-Manning}\\\text{equation}\end{array}\right] \qquad \text{19.12(b)}$$

$$Q = vA = \left(\frac{1.49}{n}\right) A R^{\frac{2}{3}} \sqrt{S}$$

$$= K\sqrt{S} \qquad \text{19.13(b)}$$

8. HAZEN-WILLIAMS VELOCITY

$$v = 1.318 C R^{0.63} S_0^{0.54} \qquad \text{19.14(b)}$$

9. NORMAL DEPTH

$$d_n = 0.788 \left(\frac{nQ}{w\sqrt{S}}\right)^{\frac{3}{5}} \quad [w \gg d_n] \qquad \text{19.15(b)}$$

$$D = d_n = 1.335 \left(\frac{nQ}{\sqrt{S}}\right)^{\frac{3}{8}} \quad \text{[full]} \qquad \text{19.16(b)}$$

$$D = 2d_n = 1.731 \left(\frac{nQ}{\sqrt{S}}\right)^{\frac{3}{8}} \quad \text{[half full]} \qquad \text{19.17(b)}$$

$$R = \frac{wd_n}{w + 2d_n} \qquad \text{19.18}$$

$$A = wd_n \qquad \text{19.19}$$

$$Q = \left(\frac{1.49}{n}\right)(wd_n)\left(\frac{wd_n}{w + 2d_n}\right)^{\frac{2}{3}}\sqrt{S}$$
$$\text{[rectangular]} \qquad \text{19.20(b)}$$

11. SIZING TRAPEZOIDAL AND RECTANGULAR CHANNELS

$$Q = \frac{K'b^{\frac{8}{3}}\sqrt{S_0}}{n} \qquad \text{19.31}$$

$$K' = \left(\frac{1.49\left(1 + m\left(\frac{d}{b}\right)\right)^{\frac{5}{3}}}{\left(1 + 2\left(\frac{d}{b}\right)\sqrt{1 + m^2}\right)^{\frac{2}{3}}}\right)\left(\frac{d}{b}\right)^{\frac{5}{3}} \qquad \text{19.32(b)}$$

12. MOST EFFICIENT CROSS SECTION

- *rectangle*

$$d = \frac{w}{2} \quad \text{[most efficient rectangle]} \qquad \text{19.34}$$

$$A = dw = \frac{w^2}{2} = 2d^2 \qquad \text{19.35}$$

$$P = d + w + d = 2w = 4d \qquad \text{19.36}$$

$$R = \frac{w}{4} = \frac{d}{2} \qquad \text{19.37}$$

- *trapezoid*

$$d = 2R \quad \text{[most efficient trapezoid]} \qquad \text{19.38}$$

$$b = \frac{2d}{\sqrt{3}} \qquad \text{19.39}$$

$$A = \sqrt{3}d^2 \qquad \text{19.40}$$

$$P = 3b = 2\sqrt{3}d \quad \text{[most efficient]} \qquad \text{19.41}$$

$$R = \frac{d}{2} \qquad \text{19.42}$$

14. FLOW MEASUREMENT WITH WEIRS

$$Q = \tfrac{2}{3}b\sqrt{2g}\left(\left(H+\frac{v_1^2}{2g}\right)^{\frac{3}{2}} - \left(\frac{v_1^2}{2g}\right)^{\frac{3}{2}}\right)$$

$$\left[\begin{array}{c}\text{rectangular}\\\text{weirs}\end{array}\right] \quad 19.47$$

$$Q = \tfrac{2}{3}b\sqrt{2g}H^{\frac{3}{2}} \quad \left[\begin{array}{c}\text{rectangular}\\\text{weirs}\end{array}\right] \quad 19.48$$

$$Q = \tfrac{2}{3}C_1 b\sqrt{2g}H^{\frac{3}{2}} \quad \left[\begin{array}{c}\text{rectangular}\\\text{weirs}\end{array}\right] \quad 19.49$$

$$C_1 = \left(0.6035 + 0.0813\left(\frac{H}{Y}\right) + \frac{0.000295}{Y}\right)$$

$$\times \left(1 + \frac{0.00361}{H}\right)^{\frac{3}{2}}$$

$$\approx 0.602 + 0.083\left(\frac{H}{Y}\right) \quad 19.50$$

$$Q \approx 3.33bh^{\frac{3}{2}} \quad \left[\frac{H}{Y} < 0.2\right] \quad 19.51(b)$$

$$b_{\text{effective}} = b_{\text{actual}} - 0.1NH \quad 19.52$$

17. BROAD-CRESTED WEIRS AND SPILLWAYS

$$Q = \tfrac{2}{3}C_1 b\sqrt{2g}H^{\frac{3}{2}} \quad 19.58$$

$$Q = C_{\text{Horton}}b\left(H+\frac{v^2}{2g}\right)^{\frac{3}{2}} \quad 19.59$$

$$Q = C_s b H^{\frac{3}{2}} \quad 19.60$$

19. FLOW MEASUREMENT WITH PARSHALL FLUMES

$$Q = KbH_a^n \quad 19.63$$

$$n = 1.522b^{0.026} \quad 19.64$$

21. SPECIFIC ENERGY

$$E = d + \frac{v^2}{2g} \quad 19.66$$

$$E = d + \frac{Q^2}{2gA^2} \quad \text{[general]} \quad 19.67$$

$$E = d + \frac{Q^2}{2g(wd)^2} \quad \text{[rectangular]} \quad 19.69$$

25. CRITICAL FLOW AND CRITICAL DEPTH IN RECTANGULAR CHANNELS

$$d_c^3 = \frac{Q^2}{gw^2} \quad \text{[rectangular]} \quad 19.74$$

$$d_c = \tfrac{2}{3}E_c \quad 19.75$$

$$v_c = \sqrt{gd_c} \quad 19.76$$

26. CRITICAL FLOW AND CRITICAL DEPTH IN NONRECTANGULAR CHANNELS

$$\frac{Q^2}{g} = \frac{A^3}{T} \quad \text{[nonrectangular]} \quad 19.77$$

27. FROUDE NUMBER

$$\text{Fr} = \frac{v}{\sqrt{gD_h}} \quad 19.78$$

33. HYDRAULIC JUMP

$$d_1 = -\tfrac{1}{2}d_2 + \sqrt{\frac{2v_2^2 d_2}{g} + \frac{d_2^2}{4}} \quad \left[\begin{array}{c}\text{rectangular}\\\text{channels}\end{array}\right] \quad 19.90$$

$$d_2 = -\tfrac{1}{2}d_1 + \sqrt{\frac{2v_1^2 d_1}{g} + \frac{d_1^2}{4}} \quad \left[\begin{array}{c}\text{rectangular}\\\text{channels}\end{array}\right] \quad 19.91$$

$$\frac{d_2}{d_1} = \tfrac{1}{2}\left(\sqrt{1+8(\text{Fr}_1)^2} - 1\right) \quad \left[\begin{array}{c}\text{rectangular}\\\text{channels}\end{array}\right] \quad 19.92(a)$$

$$\frac{d_1}{d_2} = \tfrac{1}{2}\left(\sqrt{1+8(\text{Fr}_2)^2} - 1\right) \quad \left[\begin{array}{c}\text{rectangular}\\\text{channels}\end{array}\right] \quad 19.92(b)$$

$$v_1^2 = \left(\frac{gd_2}{2d_1}\right)(d_1 + d_2) \quad \left[\begin{array}{c}\text{rectangular}\\\text{channels}\end{array}\right] \quad 19.93$$

$$\Delta E = \left(d_1 + \frac{v_1^2}{2g}\right) - \left(d_2 + \frac{v_2^2}{2g}\right)$$

$$\approx \frac{(d_2 - d_1)^3}{4d_1 d_2} \quad 19.94$$

38. DETERMINING TYPE OF CULVERT FLOW

A. Type-1 Flow

$$Q = C_d A_c \sqrt{2g\left(h_1 - z + \frac{\alpha v_1^2}{2g} - d_c - h_{f,1\text{-}2}\right)} \quad 19.101$$

B. Type-2 Flow

$$Q = C_d A_c \sqrt{2g \left(h_1 + \frac{\alpha v_1^2}{2g} - d_c - h_{f,1\text{-}2} - h_{f,2\text{-}3} \right)}$$

19.102

C. Type-3 Flow

$$Q = C_d A_3 \sqrt{2g \left(h_1 + \frac{\alpha v_1^2}{2g} - h_3 - h_{f,1\text{-}2} - h_{f,2\text{-}3} \right)}$$

19.103

D. Type-4 Flow

$$Q = C_d A_o \sqrt{2g \left(\frac{h_1 - h_4}{1 + \dfrac{29 C_d^2 n^2 L}{R^{\frac{4}{3}}}} \right)}$$

19.104

$$Q = C_d A_o \sqrt{2g(h_1 - h_4)}$$

19.105

E. Type-5 Flow

$$Q = C_d A_o \sqrt{2g(h_1 - z)}$$

19.106

F. Type-6 Flow

$$Q = C_d A_o \sqrt{2g(h_1 - h_3 - h_{f,2\text{-}3})}$$

19.107

CERM Chapter 20
Meteorology, Climatology, and Hydrology

> Chapter, section, equation, figure, and table numbers correspond to CERM. For additional study material, go to the corresponding chapter and section number in CERM.

5. TIME OF CONCENTRATION

$$t_c = t_{\text{sheet}} + t_{\text{shallow}} + t_{\text{channel}}$$

20.5

6. RAINFALL INTENSITY

$$I = \frac{K}{t_c + b}$$

20.14

7. FLOODS

$$p\{F \text{ event in } n \text{ years}\} = 1 - \left(1 - \frac{1}{F} \right)^n$$

20.20

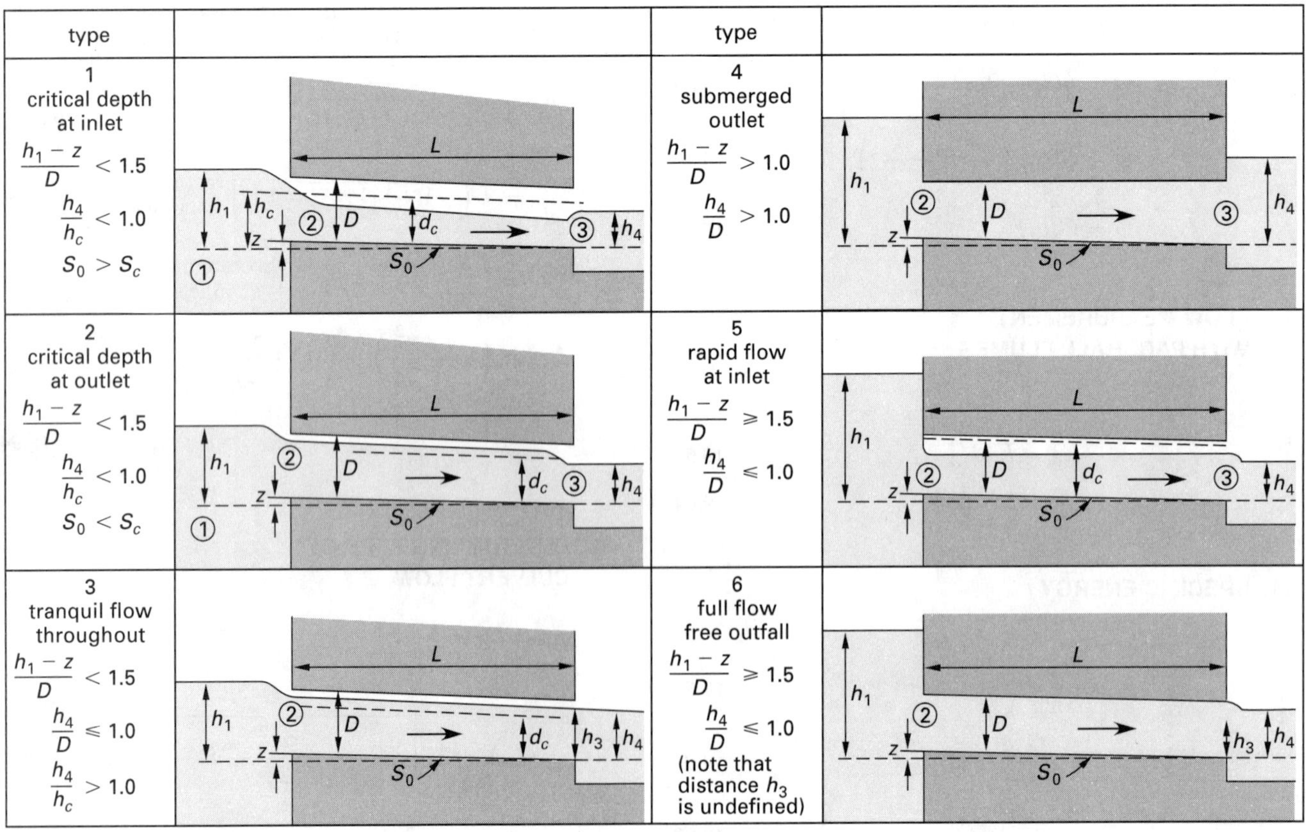

Figure 19.25 *Culvert Flow Classifications*

10. UNIT HYDROGRAPH

$$V = A_d P_{\text{ave,excess}} \qquad 20.21$$

15. PEAK RUNOFF FROM THE RATIONAL METHOD

$$Q_p = CIA_d \qquad 20.36$$

CERM Chapter 21
Groundwater

Chapter, section, equation, figure, and table numbers correspond to CERM. For additional study material, go to the corresponding chapter and section number in CERM.

2. AQUIFER CHARACTERISTICS

$$w = \frac{m_w}{m_s} = \frac{m_t - m_s}{m_s} \qquad 21.2$$

$$n = \frac{V_v}{V_t} = \frac{V_t - V_s}{V_t} \qquad 21.3$$

$$e = \frac{V_v}{V_s} = \frac{V_t - V_s}{V_s} \qquad 21.4$$

$$e = \frac{n}{1 - n} \qquad 21.5$$

$$i = \frac{\Delta H}{L} \qquad 21.6$$

3. PERMEABILITY

$$K = \frac{k\gamma}{\mu} \qquad 21.7$$

$$K = CD_{\text{mean}}^2 \qquad 21.8$$

$$K_{\text{cm/s}} \approx C(D_{10,\text{mm}})^2$$
$$[0.1 \text{ mm} \leq D_{10,\text{mm}} \leq 3.0 \text{ mm}] \qquad 21.9$$

4. DARCY'S LAW

$$Q = -KiA_{\text{gross}} = -v_e A_{\text{gross}} \qquad 21.10$$

$$q = \frac{Q}{A_{\text{gross}}} = Ki = v_e \qquad 21.11$$

$$\text{Re} = \frac{\rho q D_{\text{mean}}}{\mu} = \frac{q D_{\text{mean}}}{\nu} \qquad 21.12$$

5. TRANSMISSIVITY

$$T = KY \qquad 21.13$$

$$Q = bTi \qquad 21.14$$

6. SPECIFIC YIELD, RETENTION, AND CAPACITY

$$S_y = \frac{V_{\text{yielded}}}{V_{\text{total}}} \qquad 21.15$$

$$S_r = \frac{V_{\text{retained}}}{V_{\text{total}}} = n - S_y \qquad 21.16$$

$$\text{specific capacity} = \frac{Q}{s} \qquad 21.17$$

7. DISCHARGE VELOCITY AND SEEPAGE VELOCITY

$$v_{\text{pore}} = \frac{Q}{A_{\text{net}}} = \frac{Q}{nA_{\text{gross}}} = \frac{Q}{nbY} = \frac{Ki}{n} \qquad 21.18$$

$$v_e = nv_{\text{pore}} = \frac{Q}{A_{\text{gross}}} = \frac{Q}{bY} = Ki \qquad 21.19$$

11. WELL DRAWDOWN IN AQUIFERS

$$Q = \frac{\pi K(y_1^2 - y_2^2)}{\ln\left(\dfrac{r_1}{r_2}\right)} \quad \begin{bmatrix} \text{Dupuit} \\ \text{equation} \end{bmatrix} \qquad 21.25$$

$$Q = \frac{2\pi T(s_2 - s_1)}{\ln\left(\dfrac{r_1}{r_2}\right)} \quad [s \ll Y] \qquad 21.26$$

$$Q = \frac{2\pi KY(y_1 - y_2)}{\ln\left(\dfrac{r_1}{r_2}\right)} \quad \begin{bmatrix} \text{Thiem equation;} \\ \text{artesian well} \end{bmatrix} \qquad 21.27$$

12. UNSTEADY FLOW

$$s_{r,t} = \left(\frac{Q}{4\pi KY}\right) W(u)$$

$$= \left(\frac{Q}{4\pi T}\right) W(u) \quad [\text{Theis equation}] \qquad 21.28$$

$$s_1 - s_2 = y_2 - y_1$$

$$= \left(\frac{Q}{4\pi KY}\right) (W(u_1) - W(u_2)) \qquad 21.31$$

15. SEEPAGE FROM FLOW NETS

$$Q = KH\left(\frac{N_f}{N_p}\right) \quad [\text{per unit width}] \qquad 21.33$$

$$H = H_1 - H_2 \qquad 21.34$$

16. HYDROSTATIC PRESSURE ALONG FLOW PATH

$$p_u = \left(\frac{j}{N_p}\right) H \gamma_w \qquad 21.35$$

$$p_u = \left(\left(\frac{j}{N_p}\right) H + z\right) \gamma_w \qquad 21.36$$

$$U = N p_u A \quad \text{[uplift force]} \qquad 21.37$$

$$(\text{FS})_{\text{heave}} = \frac{\text{downward pressure}}{\text{uplift pressure}} \qquad 21.38$$

17. INFILTRATION

$$f_t = f_c + (f_0 - f_c)e^{-kt} \qquad 21.39$$

$$F_t = f_c t + \frac{f_0 - f_c}{k} \qquad 21.40$$

CERM Chapter 22
Inorganic Chemistry

> Chapter, section, equation, figure, and table numbers correspond to CERM. For additional study material, go to the corresponding chapter and section number in CERM.

9. EQUIVALENT WEIGHT

$$\text{EW} = \frac{\text{MW}}{\Delta \text{ oxidation number}} \qquad 22.2$$

10. GRAVIMETRIC FRACTION

$$x_i = \frac{m_i}{m_1 + m_2 + \cdots + m_i + \cdots + m_n} = \frac{m_i}{m_t} \qquad 22.3$$

18. UNITS OF CONCENTRATION

F— *formality:* The number of gram formula weights (i.e., molecular weights in grams) per liter of solution.

m— *molality:* The number of gram-moles of solute per 1000 grams of solvent. A "molal" solution contains 1 gram-mole per 1000 grams of solvent.

M— *molarity:* The number of gram-moles of solute per liter of solution. A "molar" (i.e., 1 M) solution contains 1 gram-mole per liter of solution. Molarity is related to normality: $N = M \times \Delta$ oxidation number.

N— *normality:* The number of gram equivalent weights of solute per liter of solution. A solution is "normal" (i.e., 1 N) if there is exactly one gram equivalent weight per liter of solution.

x— *mole fraction:* The number of moles of solute divided by the number of moles of solvent and all solutes.

meq/L— *milligram equivalent weights* of solute *per liter* of solution: calculated by multiplying normality by 1000 or dividing concentration in mg/L by equivalent weight.

mg/L— *milligrams per liter:* The number of milligrams of solute per liter of solution. Same as ppm for solutions of water.

ppm— *parts per million:* The number of pounds (or grams) of solute per million pounds (or grams) of solution. Same as mg/L for solutions of water.

19. pH AND pOH

$$\text{pH} = -\log_{10}[\text{H}^+] = \log_{10}\left(\frac{1}{[\text{H}^+]}\right) \qquad 22.8$$

$$\text{pOH} = -\log_{10}[\text{OH}^-] = \log_{10}\left(\frac{1}{[\text{OH}^-]}\right) \qquad 22.9$$

$$[X] = (\text{fraction ionized}) \times M \qquad 22.10$$

$$\text{pH} + \text{pOH} = 14 \qquad 22.11$$

26. REVERSIBLE REACTION KINETICS

$$aA + bB \rightleftharpoons cC + dD \qquad 22.15$$

$$v_{\text{forward}} = k_{\text{forward}}[A]^a[B]^b \qquad 22.16$$

$$v_{\text{reverse}} = k_{\text{reverse}}[C]^c[D]^d \qquad 22.17$$

27. EQUILIBRIUM CONSTANT

$$K = \frac{[C]^c[D]^d}{[A]^a[B]^b} = \frac{k_{\text{forward}}}{k_{\text{reverse}}} \qquad 22.19$$

28. IONIZATION CONSTANT

$$K_{\text{ionization}} = \frac{MX^2}{1 - X} \qquad 22.24$$

32. ENTHALPY OF REACTION

$$\Delta H_r = \sum \Delta H_{f,\text{products}} - \sum \Delta H_{f,\text{reactants}} \qquad 22.26$$

Table of Relative Atomic Weights
(based on the atomic mass of $^{12}C = 12$)

name	symbol	atomic number	atomic weight	name	symbol	atomic number	atomic weight
actinium	Ac	89	–	mercury	Hg	80	200.59
aluminum	Al	13	26.9815	molybdenum	Mo	42	95.94
americium	Am	95	–	neodymium	Nd	60	144.24
antimony	Sb	51	121.75	neon	Ne	10	20.183
argon	Ar	18	39.948	neptunium	Np	93	–
arsenic	As	33	74.9216	nickel	Ni	28	58.71
astatine	At	85	–	niobium	Nb	41	92.906
barium	Ba	56	137.34	nitrogen	N	7	14.0067
berkelium	Bk	97	–	nobelium	No	102	–
beryllium	Be	4	9.0122	osmium	Os	76	190.2
bismuth	Bi	83	208.980	oxygen	O	8	15.9994
boron	B	5	10.811	palladium	Pd	46	106.4
bromine	Br	35	79.904	phosphorus	P	15	30.9738
cadmium	Cd	48	112.40	platinum	Pt	78	195.09
calcium	Ca	20	40.08	plutonium	Pu	94	–
californium	Cf	98	–	polonium	Po	84	–
carbon	C	6	12.01115	potassium	K	19	39.102
cerium	Ce	58	140.12	praseodymium	Pr	59	140.907
cesium	Cs	55	132.905	promethium	Pm	61	–
chlorine	Cl	17	35.453	protactinium	Pa	91	–
chromium	Cr	24	51.996	radium	Ra	88	–
cobalt	Co	27	58.9332	radon	Rn	86	–
copper	Cu	29	63.546	rhenium	Re	75	186.2
curium	Cm	96	–	rhodium	Rh	45	102.905
dysprosium	Dy	66	162.50	rubidium	Rb	37	85.47
einsteinium	Es	99	–	ruthenium	Ru	44	101.07
erbium	Er	68	167.26	samarium	Sm	62	150.35
europium	Eu	63	151.96	scandium	Sc	21	44.956
fermium	Fm	100	–	selenium	Se	34	78.96
fluorine	F	9	18.9984	silicon	Si	14	28.086
francium	Fr	87	–	silver	Ag	47	107.868
gadolinium	Gd	64	157.25	sodium	Na	11	22.9898
gallium	Ga	31	69.72	strontium	Sr	38	87.62
germanium	Ge	32	72.59	sulfur	S	16	32.064
gold	Au	79	196.967	tantalum	Ta	73	180.948
hafnium	Hf	72	178.49	technetium	Tc	43	–
helium	He	2	4.0026	tellurium	Te	52	127.60
holmium	Ho	67	164.930	terbium	Tb	65	158.924
hydrogen	H	1	1.00797	thallium	Tl	81	204.37
indium	In	49	114.82	thorium	Th	90	232.038
iodine	I	53	126.9044	thulium	Tm	69	168.934
iridium	Ir	77	192.2	tin	Sn	50	118.69
iron	Fe	26	55.847	titanium	Ti	22	47.90
krypton	Kr	36	83.80	tungsten	W	74	183.85
lanthanum	La	57	138.91	uranium	U	92	238.03
lead	Pb	82	207.19	vanadium	V	23	50.942
lithium	Li	3	6.939	xenon	Xe	54	131.30
lutetium	Lu	71	174.97	ytterbium	Yb	70	173.04
magnesium	Mg	12	24.312	yttrium	Y	39	88.905
manganese	Mn	25	54.9380	zinc	Zn	30	65.37
mendelevium	Md	101	–	zirconium	Zr	40	91.22

CERM Chapter 24
Combustion and Incineration

Chapter, section, equation, figure, and table numbers correspond to CERM. For additional study material, go to the corresponding chapter and section number in CERM.

6. MOISTURE

$$G_{H,\text{combined}} = \frac{G_O}{8} \qquad 24.1$$

$$G_{H,\text{available}} = G_{H,\text{total}} - \frac{G_O}{8} \qquad 24.2$$

24. ATMOSPHERIC AIR

Table 24.6 Composition of Dry Air [a]

component	percent by weight	percent by volume
oxygen	23.15	20.95
nitrogen/inerts	76.85	79.05
ratio of nitrogen to oxygen	3.320	3.773[b]
ratio of air to oxygen	4.320	4.773

[a] Inert gases and CO_2 included as N_2.
[b] The value is also reported by various sources as 3.76, 3.78, and 3.784.

25. COMBUSTION REACTIONS

Table 24.7 Ideal Combustion Reactions

fuel	formula	reaction equation (excluding nitrogen)
carbon (to CO)	C	$2C + O_2 \longrightarrow 2CO$
carbon (to CO_2)	C	$C + O_2 \longrightarrow CO_2$
sulfur (to SO_2)	S	$S + O_2 \longrightarrow SO_2$
sulfur (to SO_3)	S	$2S + 3O_2 \longrightarrow 2SO_3$
carbon monoxide	CO	$2CO + O_2 \longrightarrow 2CO_2$
methane	CH_4	$CH_4 + 2O_2 \longrightarrow CO_2 + 2H_2O$
acetylene	C_2H_2	$2C_2H_2 + 5O_2 \longrightarrow 4CO_2 + 2H_2O$
ethylene	C_2H_4	$C_2H_4 + 3O_2 \longrightarrow 2CO_2 + 2H_2O$
ethane	C_2H_6	$2C_2H_6 + 7O_2 \longrightarrow 4CO_2 + 6H_2O$
hydrogen	H_2	$2H_2 + O_2 \longrightarrow 2H_2O$
hydrogen sulfide	H_2S	$2H_2S + 3O_2 \longrightarrow 2H_2O + 2SO_2$
propane	C_3H_8	$C_3H_8 + 5O_2 \longrightarrow 3CO_2 + 4H_2O$
n-butane	C_4H_{10}	$2C_4H_{10} + 13O_2 \longrightarrow 8CO_2 + 10H_2O$
octane	C_8H_{18}	$2C_8H_{18} + 25O_2 \longrightarrow 16CO_2 + 18H_2O$
olefin series	C_nH_{2n}	$2C_nH_{2n} + 3nO_2 \longrightarrow 2nCO_2 + 2nH_2O$
paraffin series	C_nH_{2n+2}	$2C_nH_{2n+2} + (3n+1)O_2 \longrightarrow 2nCO_2 + (2n+2)H_2O$

(Multiply oxygen volumes by 3.773 to get nitrogen volumes.)

27. STOICHIOMETRIC AIR

$$R_{a/f,\text{ideal}} = \frac{m_{\text{air,ideal}}}{m_{\text{fuel}}} \qquad 24.5$$

$$R_{a/f,\text{ideal}} = (34.5)\left(\frac{G_C}{3} + G_H - \frac{G_O}{8} + \frac{G_S}{8}\right)$$
$$\text{[solid fuels]} \qquad 24.6$$

$$R_{a/f,\text{ideal}} = \sum J_i G_i \quad \text{[gaseous fuels]} \qquad 24.7$$

$$\frac{\text{volumetric air/}}{\text{fuel ratio}} = \sum K_i B_i \quad \text{[gaseous fuels]} \qquad 24.8$$

30. ACTUAL AND EXCESS AIR

$$R_{a/f,\text{actual}} = \frac{m_{\text{air,actual}}}{m_{\text{fuel}}}$$

$$= \frac{3.04 B_{N_2}\left(G_C + \dfrac{G_S}{1.833}\right)}{B_{CO_2} + B_{CO}} \qquad 24.9$$

34. DEW POINT OF FLUE GAS MOISTURE

$$\text{partial pressure} = (\text{water vapor mole fraction})$$
$$\times (\text{flue gas pressure}) \qquad 24.13$$

35. HEAT OF COMBUSTION

$$\text{HHV} = \text{LHV} + m_{\text{water}} h_{fg} \qquad 24.14$$

$$G_{H,\text{available}} = G_{H,\text{total}} - \frac{G_O}{8} \qquad 24.15$$

$$\text{HHV}_{\text{Btu/lbm}} = 14{,}093 G_C + 60{,}958\left(G_H - \frac{G_O}{8}\right)$$
$$+ 3983 G_S \quad [\text{solid fuels}] \qquad 24.16(b)$$

$$\text{HHV}_{\text{gasoline,Btu/lbm}} = 18{,}320$$
$$+ 40(^\circ\text{Baumé} - 10) \qquad 24.17(b)$$

$$\text{HHV}_{\text{fuel oil,Btu/lbm}} = 22{,}320 - 3780(\text{SG})^2 \qquad 24.18(b)$$

36. MAXIMUM THEORETICAL COMBUSTION (FLAME) TEMPERATURE

$$T_{\max} = T_i + \frac{\text{lower heat of combustion}}{m_{\text{products}} c_{p,\text{mean}}} \qquad 24.19$$

37. COMBUSTION LOSSES

$$q_1 = m_{\text{flue gas}} c_p (T_{\text{flue gas}} - T_{\text{incoming air}}) \qquad 24.20$$

$$q_2 = m_{\text{vapor}}(h_g - h_f) = 8.94 G_H(h_g - h_f) \qquad 24.21$$

$$q_3 = m_{\text{atmospheric water vapor}}(h_g - h_g')$$
$$= \omega m_{\text{combustion air}}(h_g - h_g') \qquad 24.22$$

$$q_4 = \frac{(\text{HHV}_C - \text{HHV}_{CO}) G_C B_{CO}}{B_{CO_2} + B_{CO}} \qquad 24.23$$

$$q_5 = \text{HHV}_C m_{\text{ash}} G_{C,\text{ash}} \qquad 24.24$$

38. COMBUSTION EFFICIENCY

$$\eta = \frac{\text{useful heat extracted}}{\text{heating value}}$$

$$= \frac{m_{\text{steam}}(h_{\text{steam}} - h_{\text{feedwater}})}{\text{HHV}} \qquad 24.25$$

$$\eta = \frac{\text{HHV} - q_1 - q_2 - q_3 - q_4 - q_5 - \text{radiation}}{\text{HHV}}$$

$$= \frac{\text{LHV} - q_1 - q_4 - q_5 - \text{radiation}}{\text{HHV}} \qquad 24.26$$

CERM Chapter 25
Water Supply Quality and Testing

Chapter, section, equation, figure, and table numbers correspond to CERM. For additional study material, go to the corresponding chapter and section number in CERM.

6. HARDNESS AND ALKALINITY

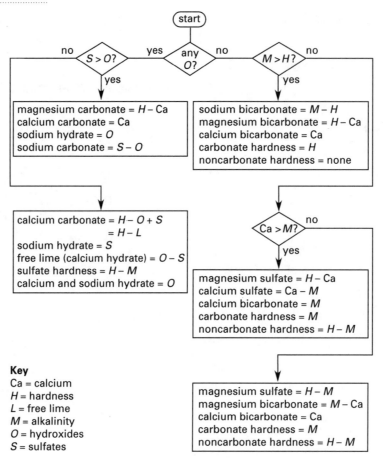

Key
Ca = calcium
H = hardness
L = free lime
M = alkalinity
O = hydroxides
S = sulfates

[a] Refer to the chapter nomenclature.
[b] All concentrations are expressed as $CaCO_3$.
[c] Not for use when other ionic species are present in significant quantities.

Figure 25.1 Hardness and Alkalinity[a,b,c]

Conversions From mg/L as a Substance to mg/L as CaCO₃

Multiply the mg/L of the substances listed in the following table by the corresponding factors to obtain mg/L as $CaCO_3$. For example, 70 mg/L of Mg^{++} would be (70 mg/L)(4.10) = 287 mg/L as $CaCO_3$.

substance	factor	substance	factor
Al^{+++}	5.56	HCO_3^-	0.82
$Al_2(SO_4)_3$	0.88^a	K^+	1.28
$AlCl_3$	1.13	KCl	0.67
$Al(OH)_3$	1.92	K_2CO_3	0.72
Ba^{++}	0.73	Mg^{++}	4.10
$Ba(OH)_2$	0.59	$MgCl_2$	1.05
$BaSO_4$	0.43	$MgCO_3$	1.19
Ca^{++}	2.50	$Mg(HCO_3)_2$	0.68
$CaCl_2$	0.90	MgO	2.48
$CaCO_3$	1.00	$Mg(OH)_2$	1.71
$Ca(HCO_3)_2$	0.62	$Mg(NO_3)_2$	0.67
CaO	1.79	$MgSO_4$	0.83
$Ca(OH)_2$	1.35	Mn^{++}	1.82
$CaSO_4$	0.74^a	Na^+	2.18
Cl^-	1.41	$NaCl$	0.85
CO_2	2.27	Na_2CO_3	0.94
CO_3^{--}	1.67	$NaHCO_3$	0.60
Cu^{++}	1.57	$NaNO_3$	0.59
Cu^{+++}	2.36	$NaOH$	1.25
$CuSO_4$	0.63	Na_2SO_4	0.70^a
F^-	2.66	NH_3	2.94
Fe^{++}	1.79	NH_4^+	2.78
Fe^{+++}	2.69	NH_4OH	1.43
$Fe(OH)_3$	1.41	$(NH_4)_2SO_4$	0.76
$FeSO_4$	0.66^a	NO_3^-	0.81
$Fe_2(SO_4)_3$	0.75	OH^-	2.94
$FeCl_3$	0.93	PO_4^{---}	1.58
H^+	50.0	SO_4^{--}	1.04
		Zn^{++}	1.54

^{a}anhydrous

CERM Chapter 26
Water Supply Treatment and Distribution

Chapter, section, equation, figure, and table numbers correspond to CERM. For additional study material, go to the corresponding chapter and section number in CERM.

11. SEDIMENTATION PHYSICS

$$\mathrm{Re} = \frac{v_s D}{\nu} \qquad 26.2$$

$$v_{s,\text{ft/sec}} = \frac{(\rho_{\text{particle}} - \rho_{\text{water}})D_{\text{ft}}^2 g}{18\mu g_c} \quad \begin{bmatrix} \text{settling} \\ \text{velocity} \end{bmatrix}$$

$$= \frac{(\text{SG}_{\text{particle}} - 1)D_{\text{ft}}^2 g}{18\nu} \quad \text{[Stokes' law]} \quad 26.3(b)$$

$$v_s = \sqrt{\frac{4gD(\text{SG}_{\text{particle}} - 1)}{3C_D}} \qquad 26.4$$

12. SEDIMENTATION TANKS

$$t_{\text{settling}} = \frac{h}{v_s} \qquad 26.5$$

$$t_d = \frac{V_{\text{tank}}}{Q} = \frac{Ah}{Q} \qquad 26.6$$

$$v^* = \frac{Q_{\text{filter}}}{A_{\text{surface}}} = \frac{Q_{\text{filter}}}{bL} \qquad 26.7$$

15. DOSES OF COAGULANTS AND OTHER COMPOUNDS

$$F_{\text{lbm/day}} = \frac{D_{\text{mg/L}}Q_{\text{MGD}}\left(8.345 \ \dfrac{\text{lbm-L}}{\text{mg·MG}}\right)}{PG} \qquad 26.11(b)$$

16. MIXERS AND MIXING KINETICS

$$V = tQ \qquad 26.12$$

$$t_{\text{complete}} = \frac{V}{Q} = \frac{1}{K}\left(\frac{C_i}{C_o} - 1\right) \qquad 26.13$$

$$t_{\text{plug flow}} = \frac{V}{Q}$$

$$= \frac{L}{v_{\text{flow through}}} = \frac{1}{K}\ln\left(\frac{C_i}{C_o}\right) \qquad 26.14$$

17. MIXING PHYSICS

$$F_D = \frac{C_D A \rho v_{mixing}^2}{2g_c} = \frac{C_D A \gamma v_{mixing}^2}{2g} \qquad 26.15(b)$$

$$v_{paddle,ft/sec} = \frac{2\pi R n_{rpm}}{60 \frac{sec}{min}} \qquad 26.16$$

$$v_{mixing} = v_{paddle} - v_{water} \qquad 26.17$$

$$P_{hp} = \frac{F_D v_{mixing}}{550 \frac{ft\text{-}lbf}{hp\text{-}sec}}$$

$$= \frac{C_D A \gamma v_{mixing}^3}{2g \left(550 \frac{ft\text{-}lbf}{hp\text{-}sec}\right)} \qquad 26.19(b)$$

$$G = \sqrt{\frac{P}{\mu V_{tank}}} \qquad 26.20$$

$$P = \mu G^2 V_{tank} \qquad 26.21$$

$$Gt_d = \frac{V_{tank}}{G}\sqrt{\frac{P}{\mu V_{tank}}}$$

$$= \frac{1}{Q}\sqrt{\frac{P V_{tank}}{\mu}} \qquad 26.22$$

18. IMPELLER CHARACTERISTICS

$$Re = \frac{D^2 n \rho}{g_c \mu} \qquad 26.23(b)$$

$$P = \frac{N_P n^3 D^5 \rho}{g_c} \quad \begin{bmatrix} \text{geometrically} \\ \text{similar; turbulent} \end{bmatrix} \quad 26.24(b)$$

$$P = \frac{\rho g Q h_v}{g_c} \quad \begin{bmatrix} \text{geometrically} \\ \text{similar; turbulent} \end{bmatrix} \quad 26.25(b)$$

$$Q = N_Q n D^3 \quad \begin{bmatrix} \text{geometrically} \\ \text{similar; turbulent} \end{bmatrix} \quad 26.26$$

$$h_v = \frac{N_P n^2 D^2}{N_Q g} \quad \begin{bmatrix} \text{geometrically} \\ \text{similar; turbulent} \end{bmatrix} \quad 26.27$$

21. SLUDGE QUANTITIES

$$V_{sludge} = \frac{m_{sludge}}{\rho_{water}(SG)_{sludge} G} \qquad 26.28$$

$$m_{sludge,lbm/day} = \left(8.345 \frac{lbm\text{-}L}{mg\text{-}MG}\right)(Q_{MGD})$$
$$\times \left(\begin{array}{c} 0.46 D_{alum,mg/L} \\ + \Delta TSS_{mg/L} + M_{mg/L} \end{array}\right) \quad 26.29(b)$$

$$V_{2,sludge} \approx V_{1,sludge}\left(\frac{G_1}{G_2}\right) \qquad 26.30$$

22. FILTRATION

$$\text{loading rate} = \frac{Q}{A} \qquad 26.31$$

23. FILTER BACKWASHING

$$V = A_{filter}(\text{rate of rise})t_{backwash} \qquad 26.32$$

26. FLUORIDATION

$$F_{lbm/day} = \frac{C_{mg/L} Q_{MGD}\left(8.345 \frac{lbm\text{-}L}{mg\text{-}MG}\right)}{PG} \qquad 26.33(b)$$

28. TASTE AND ODOR CONTROL

$$\text{TON} = \frac{V_{raw\ sample} + V_{dilution\ water}}{V_{raw\ sample}}$$

31. WATER SOFTENING BY ION EXCHANGE

$$V_{water} = \frac{(\text{specific working capacity})(V_{exchange\ material})}{\text{hardness}}$$
$$26.42$$

32. REGENERATION OF ION EXCHANGE RESINS

$$m_{salt} = (\text{specific working capacity})(V_{exchange\ material})$$
$$\times (\text{salt requirement}) \qquad 26.43$$

33. STABILIZATION

$$\text{LI} = \text{pH} - \text{pH}_{sat} \qquad 26.44$$

$$\text{pH}_{sat} = (\text{pK}_2 - \text{pK}_1) + \text{pCa} + \text{pM}$$
$$= -\log\left(\frac{K_2}{K_1}\right)[Ca^{++}][M] \qquad 26.45$$

$$\text{RI} = 2\text{pH}_{sat} - \text{pH} \qquad 26.46$$

40. WATER DEMAND

$$Q_{instantaneous} = M(\text{AADF}) \qquad 26.51$$

41. FIRE FIGHTING DEMAND

$$Q_{gpm} = 18F\sqrt{A_{ft^2}} \qquad 26.52$$

$$Q_{gpm} = 1020\sqrt{P}\left(1 - 0.01\sqrt{P}\right) \qquad 26.53$$

Environmental

CERM Chapter 28
Wastewater Quantity and Quality

Chapter, section, equation, figure, and table numbers correspond to CERM. For additional study material, go to the corresponding chapter and section number in CERM.

4. WASTEWATER QUANTITY

$$\frac{Q_{\text{peak}}}{Q_{\text{ave}}} = \frac{18 + \sqrt{P}}{4 + \sqrt{P}} \quad \left[\begin{array}{l}\text{TSS equation;}\\ P \text{ in thousands}\end{array}\right] \qquad 28.1$$

P = Population

15. MICROBIAL GROWTH

$$r_g = \frac{dX}{dt} = \mu X \qquad 28.7$$

$$\mu = \mu_m\left(\frac{S}{K_s + S}\right) \qquad 28.8$$

$$r_g = \frac{\mu_m X S}{K_s + S} \qquad 28.9$$

$$r_{su} = \frac{dS}{dt} = \frac{-\mu_m X S}{Y(K_s + S)} \qquad 28.10$$

$$k = \frac{\mu_m}{Y} \qquad 28.11$$

$$r_d = -k_d X \qquad 28.13$$

$$r'_g = \frac{\mu_m X S}{K_s + S} - k_d X$$

$$= -Y r_{su} - k_d X \qquad 28.14$$

$$\mu' = \mu_m\left(\frac{S}{K_s + S}\right) - k_d \qquad 28.15$$

16. DISSOLVED OXYGEN IN WASTEWATER

$$D = \text{DO}_{\text{sat}} - \text{DO} \qquad 28.16$$

17. REOXYGENATION

$$r_r = K_r(\text{DO}_{\text{sat}} - \text{DO}) \qquad 28.17$$

$$D_t = D_0 10^{-K_r t} \quad [\text{base-10}] \qquad 28.18$$

$$D_t = D_0 e^{-K'_r t} \quad [\text{base-}e] \qquad 28.19$$

$$K'_r = 2.303 K_r \qquad 28.20$$

$$K'_{r,20°C} \approx \frac{3.93\sqrt{v_{\text{m/s}}}}{d_{\text{m}}^{1.5}} \quad \left[\begin{array}{l}\text{O'Connor}\\ \text{and Dobbins}\end{array}\right]$$

$$\left[\begin{array}{c}0.3\text{ m} < d < 9.14\text{ m}\\ 0.15\text{ m/s} < v < 0.49\text{ m/s}\end{array}\right]$$

28.21(a)

$$K'_{r,68°F} \approx \frac{12.9\sqrt{v_{\text{ft/sec}}}}{d_{\text{ft}}^{1.5}} \quad \left[\begin{array}{l}\text{O'Connor}\\ \text{and Dobbins}\end{array}\right]$$

$$\left[\begin{array}{c}1\text{ ft} < d < 30\text{ ft}\\ 0.5\text{ ft/sec} < v < 1.6\text{ ft/sec}\end{array}\right]$$

28.21(b)

$$K'_{r,20°C} \approx \frac{5.026\, v_{\text{m/s}}}{d_{\text{m}}^{1.67}} \quad [\text{Churchill}]$$

$$\left[\begin{array}{c}0.61\text{ m} < d < 3.35\text{ m}\\ 0.55\text{ m/s} < v < 1.52\text{ m/s}\end{array}\right] \qquad 28.22$$

$$K'_{r,T} = K'_{r,20°C}(1.024)^{T-20°C} \qquad 28.23$$

$$K'_{r,T_1} = K'_{r,T_2}\theta_r^{T_1-T_2} \qquad 28.24$$

18. DEOXYGENATION

$$r_{d,t} = -K_d\text{BOD}_u \qquad 28.25$$

$$K'_{d,T} = K'_{d,20°C}\theta^{T-20°C} \qquad 28.26$$

$$K'_{d,T_1} = K'_{d,T_2}\theta_d^{T_1-T_2} \qquad 28.27$$

20. BIOCHEMICAL OXYGEN DEMAND

$$\text{BOD}_5 = \frac{\text{DO}_i - \text{DO}_f}{\dfrac{V_{\text{sample}}}{V_{\text{sample}} + V_{\text{dilution}}}} \qquad 28.28$$

$$\text{BOD}_t = \text{BOD}_u(1 - 10^{-K_d t}) \qquad 28.29$$

$$\text{BOD}_u \approx 1.463\text{BOD}_5 \qquad 28.30$$

$$\text{BOD}_{T°C} = \text{BOD}_{20°C}(0.02T°C + 0.6) \qquad 28.31$$

21. SEEDED BOD

$$\text{BOD} = \frac{\text{DO}_i - \text{DO}_f - x(\text{DO}_i^* - \text{DO}_f^*)}{\dfrac{V_{\text{sample}}}{V_{\text{sample}} + V_{\text{dilution}}}} \qquad 28.32$$

22. DILUTION PURIFICATION

$$C_f = \frac{C_1 Q_1 + C_2 Q_2}{Q_1 + Q_2} \qquad 28.33$$

23. RESPONSE TO DILUTION PURIFICATION

$$D_t = \left(\frac{K_d \text{BOD}_u}{K_r - K_d}\right)(10^{-K_d t} - 10^{-K_r t}) + D_0(10^{-K_r t})$$
$$\qquad 28.34$$

$$x_c = \text{v} t_c \qquad 28.35$$

$$t_c = \left(\frac{1}{K_r - K_d}\right)$$
$$\times \log_{10}\left(\left(\frac{K_d \text{BOD}_u - K_r D_0 + K_d D_0}{K_d \text{BOD}_u}\right)\left(\frac{K_r}{K_d}\right)\right)$$
$$\qquad 28.36$$

$$D_c = \left(\frac{K_d \text{BOD}_u}{K_r}\right)10^{-K_d t_c} \qquad 28.37$$

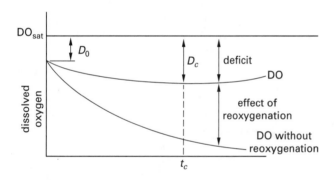

Figure 28.5 *Oxygen Sag Curve*

CERM Chapter 29
Wastewater Treatment: Equipment and Processes

Chapter, section, equation, figure, and table numbers correspond to CERM. For additional study material, go to the corresponding chapter and section number in CERM.

11. AERATED LAGOONS

$$t_d = \frac{V}{Q} = k_{1,\text{base-}e}^{1-\eta} \qquad 29.1$$

$$k_{1,\text{base-}e} = 2.3 k_{1,\text{base-}10} \qquad 29.2$$

13. GRIT CHAMBERS

$$\text{v} = \sqrt{8k\left(\frac{g d_p}{f}\right)(\text{SG}_p - 1)} \qquad 29.3$$

17. PLAIN SEDIMENTATION BASINS/CLARIFIERS

$$\text{v}^* = \frac{Q}{A} \qquad 29.4$$

$$t_d = \frac{V}{Q} \qquad 29.5$$

$$\text{weir loading} = \frac{Q}{L} \qquad 29.6$$

19. TRICKLING FILTERS

$$S_{ps} + RS_o = (1+R)S_i \qquad 29.7$$

$$S_i = \frac{S_{ps} + RS_o}{1+R} \qquad 29.8$$

$$\eta = \frac{S_{\text{removed}}}{S_{ps}} = \frac{S_{ps} - S_o}{S_{ps}} \qquad 29.9$$

$$R = \frac{Q_r}{Q_w} \qquad 29.10$$

$$L_H = \frac{Q_w + Q_r}{A} = \frac{Q_w(1+R)}{A} \qquad 29.11$$

$$L_{\text{BOD}} = \frac{Q_{w,\text{MGD}}S_{\text{mg/L}} \times \left(8.345 \dfrac{\text{lbm-L}}{\text{MG-mg}}\right)(1000)}{V_{\text{ft}^3}} \qquad 29.12$$

21. NATIONAL RESEARCH COUNCIL EQUATION

$$\eta = \frac{1}{1 + 0.0561\sqrt{\dfrac{L_{\text{BOD}}}{F}}} \qquad 29.13$$

$$\eta = \frac{1}{1 + 0.0085\sqrt{\dfrac{L_{\text{BOD,lbm/day}}}{V_{\text{ac-ft}}F}}} \qquad 29.14$$

$$F = \frac{1+R}{(1+wR)^2} \qquad 29.15$$

$$\eta_2 = \frac{1}{1 + \left(\dfrac{0.0561}{1-\eta_1}\right)\sqrt{\dfrac{L_{\text{BOD}}}{F}}} \quad \begin{bmatrix}\text{two-stage}\\\text{with clarifier}\end{bmatrix} \qquad 29.16$$

22. VELZ EQUATION

$$\frac{\Delta S_i}{S_i} = \frac{S_i - S_o}{S_i} = 10^{-KZ} \qquad 29.17$$

$$K_T = K_{20°C}(1.047)^{T-20°C} \qquad 29.18$$

CERM Chapter 30
Activated Sludge and Sludge Processing

Chapter, section, equation, figure, and table numbers correspond to CERM. For additional study material, go to the corresponding chapter and section number in CERM.

5. SLUDGE PARAMETERS

$$F = S_o Q_o \qquad 30.1$$

$$M = V_a X \qquad 30.2$$

$$\text{F:M} = \frac{S_{o,\text{mg/L}} Q_{o,\text{MGD}}}{V_{a,\text{MG}} X_{\text{mg/L}}} \qquad$$

$$= \frac{S_{o,\text{mg/L}}}{\theta_{\text{days}} X_{\text{mg/L}}} \qquad 30.3$$

$$\text{F:M} = \frac{S_{o,\text{mg/L}} Q_{o,\text{MGD}}}{V_{a,\text{MG}} \text{MLSS}_{\text{mg/L}}} \qquad 30.4$$

$$\theta_c = \frac{V_a X}{Q_e X_e + Q_w X_w} \qquad 30.5$$

$$\theta_{\text{BOD}} = \frac{1}{\text{F:M}}$$

$$= \frac{V_{a,\text{MG}} X_{\text{mg/L}}}{S_{o,\text{mg/L}} Q_{o,\text{MGD}}} \qquad 30.6$$

$$\text{SVI} = \frac{\left(1000 \, \frac{\text{mg}}{\text{g}}\right) V_{\text{settled,mL/L}}}{\text{MLSS}_{\text{mg/L}}} \qquad 30.7$$

$$\text{TSS}_{\text{mg/L}} = \frac{\left(1000 \, \frac{\text{mg}}{\text{g}}\right)\left(1000 \, \frac{\text{mL}}{\text{L}}\right)}{\text{SVI}_{\text{mL/g}}} \qquad 30.8$$

6. SOLUBLE BOD ESCAPING TREATMENT

$$\text{BOD}_e = \text{BOD}_{\text{escaping treatment}}$$
$$+ \text{BOD}_{\text{effluent suspended solids}}$$
$$= S + S_e$$
$$= S + 1.42 f G X_e \qquad 30.9$$

$$f = \frac{\text{BOD}_5}{\text{BOD}_u} \qquad 30.10$$

7. PROCESS EFFICIENCY

$$\eta_{\text{BOD}} = \frac{S_o - S}{S_o} \qquad 30.11$$

8. PLUG FLOW AND STIRRED TANK MODELS

$$\frac{1}{\theta_c} = \frac{\mu_m(S_o - S)}{S_o - S + (1+R)K_s \ln\left(\frac{S_i}{S}\right)} - k_d$$
$$\text{[PFR only]} \; 30.12$$

$$\mu_m = kY \qquad 30.13$$

$$S_i = \frac{S_o + RS}{1+R} \quad \text{[PFR only]} \qquad 30.14$$

$$r_{su} = \frac{-\mu_m S X}{Y(K_s + S)} \quad \text{[PFR only]} \qquad 30.15$$

$$U = \eta(\text{F:M}) = \frac{-r_{su}}{X} = \frac{S_o - S}{\theta X} \qquad 30.16$$

$$\frac{1}{\theta_c} = Y(\text{F:M})\eta - k_d = -Y\left(\frac{r_{su}}{X}\right) - k_d$$
$$= YU - k_d \quad \text{[CSTR only]} \qquad 30.17$$

$$S = \frac{K_s(1 + k_d\theta_c)}{\theta_c(\mu_m - k_d) - 1} \quad \text{[CSTR only]} \qquad 30.18$$

$$V_a = \theta Q_o$$
$$= \frac{\theta_c Q_o Y(S_o - S)}{X(1 + k_d\theta_c)} \quad \text{[PFR and CSTR]} \; 30.19$$

$$\theta = \frac{V_a}{Q_o} \quad \text{[PFR and CSTR]} \qquad 30.20$$

$$\theta_s = \frac{V_a + V_s}{Q_o} \quad \text{[PFR and CSTR]} \qquad 30.21$$

$$X = \frac{\left(\frac{\theta_c}{\theta}\right) Y(S_o - S)}{1 + k_d\theta_c}$$
$$\text{[PFR and CSTR]} \qquad 30.22$$

$$Y_{\text{obs}} = \frac{Y}{1 + k_d\theta_c} \quad \text{[PFR and CSTR]} \qquad 30.23$$

$$P_{x,\text{kg/day}} = \frac{Q_{w,\text{m}^3/\text{day}} X_{r,\text{mg/L}}}{1000 \, \frac{\text{g}}{\text{kg}}}$$
$$= \frac{Y_{\text{obs,mg/mg}} Q_{o,\text{m}^3/\text{day}}(S_o - S)_{\text{mg/L}}}{1000 \, \frac{\text{g}}{\text{kg}}}$$
$$\text{[PFR and CSTR]} \qquad 30.24$$

$$m_{e,\text{solids}} = Q_e X_e \quad \text{[PFR and CSTR]} \qquad 30.25$$

$$\theta_c = \frac{V_a X}{Q_w X_r + Q_e X_e}$$
$$= \frac{V_a X}{Q_w X_r + (Q_o - Q_w)X_e}$$
$$\text{[PFR and CSTR]} \qquad 30.26$$

$$\theta_c \approx \frac{V_a X}{Q_w X_r} \quad \text{[PFR and CSTR]} \qquad 30.27$$

$$\theta_c \approx \frac{V_a}{Q_w} \quad \text{[CSTR]} \qquad 30.28$$

10. AERATION TANKS

$$\theta = \frac{V_a}{Q_o} \qquad 30.29$$

$$L_{\text{BOD}} = \frac{S_o Q_o}{V_a \left(1000 \, \frac{\text{mg·m}^3}{\text{kg·L}}\right)} \qquad 30.30$$

$$L_{\text{BOD}} = \frac{S_{o,\text{mg/L}} Q_{o,\text{MGD}} \times \left(8.345 \, \frac{\text{lbm-L}}{\text{MG-mg}}\right)(1000)}{V_{a,\text{ft}^3}} \qquad 30.31$$

$$\dot{m}_{\text{oxygen}} = K_t D \qquad 30.32$$

$$D = \beta \text{DO}_{\text{saturated water}} - \text{DO}_{\text{mixed liquor}} \qquad 30.33$$

$$\dot{m}_{\text{oxygen,kg/day}} = \left(\frac{Q_{o,\text{m}^3/\text{day}}(S_o - S)_{\text{mg/L}}}{f}\right) - 1.42 P_{x,\text{kg/day}}$$
$$\text{[PFR and CSTR]} \qquad 30.34$$

$$\dot{m}_{\text{air}} = \frac{\dot{m}_{\text{oxygen}}}{0.232 \eta_{\text{transfer}}} \qquad 30.35$$

$$\dot{V}_{\text{air}} = \frac{\dot{m}_{\text{air}}}{\rho_{\text{air}}} \quad \text{[PFR and CSTR]} \qquad 30.36$$

$$\frac{\dot{V}_{\text{air}} \left(1000 \, \frac{\text{mg·m}^3}{\text{kg·L}}\right)}{Q_o(S_o - S)} \qquad 30.37$$

11. AERATION POWER AND COST

$$P_{\text{ideal,hp}} = -\left(\frac{k p_1 \dot{V}_1}{(k-1)\left(550 \, \frac{\text{ft-lbf}}{\text{hp-sec}}\right)}\right)\left(1 - \left(\frac{p_2}{p_1}\right)^{\frac{k-1}{k}}\right)$$

$$= -\left(\frac{k \dot{m} R_{\text{air}} T_1}{(k-1)\left(550 \, \frac{\text{ft-lbf}}{\text{hp-sec}}\right)}\right)\left(1 - \left(\frac{p_2}{p_1}\right)^{\frac{k-1}{k}}\right)$$

$$= -\left(\frac{c_p J \dot{m} T_1}{550 \, \frac{\text{ft-lbf}}{\text{hp-sec}}}\right)\left(1 - \left(\frac{p_2}{p_1}\right)^{\frac{k-1}{k}}\right) \qquad 30.38(b)$$

$$P_{\text{actual}} = \frac{P_{\text{ideal}}}{\eta_c} \qquad 30.39$$

$$\text{total cost} = C_{\text{kW-hr}} P_{\text{actual}} t \qquad 30.40$$

12. RETURN RATE/RECYCLE RATIO

$$R = \frac{Q_r}{Q_o} \qquad 30.41$$

$$X_r(Q_o + Q_r) = X Q_r \qquad 30.42$$

$$\frac{Q_r}{Q_o + Q_r} = \frac{X_r}{X} \qquad 30.43$$

$$\frac{Q_r}{Q_o + Q_r} = \frac{V_{\text{settled,mL/L}}}{1000 \, \frac{\text{mL}}{\text{L}}} \qquad 30.44$$

$$R = \frac{V_{\text{settled,mL/L}}}{1000 \, \frac{\text{mL}}{\text{L}} - V_{\text{settled,mL/L}}} \qquad 30.45$$

14. QUANTITIES OF SLUDGE

$$m_{\text{wet}} = V \rho_{\text{sludge}} = V(\text{SG}_{\text{sludge}}) \rho_{\text{water}} \qquad 30.46$$

$$\frac{1}{\text{SG}_{\text{sludge}}} = \frac{1-s}{1} + \frac{s}{\text{SG}_{\text{solids}}}$$

$$= \frac{1 - s_{\text{fixed}} - s_{\text{volatile}}}{1} + \frac{s_{\text{fixed}}}{\text{SG}_{\text{fixed solids}}} + \frac{s_{\text{volatile}}}{\text{SG}_{\text{volatile solids}}} \qquad 30.47$$

$$V_{\text{sludge,wet}} = \frac{m_{\text{dried}}}{s \rho_{\text{sludge}}} \approx \frac{m_{\text{dried}}}{s \rho_{\text{water}}} \qquad 30.48$$

$$m_{\text{dried}} = (\Delta \text{SS})_{\text{mg/L}} Q_{o,\text{MGD}} \left(8.345 \, \frac{\text{lbm-L}}{\text{MG-mg}}\right) \qquad 30.49$$

$$m_{\text{dried}} = K S_{o,\text{mg/L}} Q_{o,\text{MGD}} \left(8.345 \, \frac{\text{lbm-L}}{\text{MG-mg}}\right)$$

$$= Y_{\text{obs}}(S_{o,\text{mg/L}} - S) Q_{o,\text{MGD}} \left(8.345 \, \frac{\text{lbm-L}}{\text{MG-mg}}\right) \qquad 30.50$$

19. METHANE PRODUCTION

$$V_{\text{methane,ft}^3/\text{day}} = \left(5.61 \frac{\text{ft}^3}{\text{lbm}}\right)$$

$$\times \left(\begin{array}{c} (ES_{o,\text{mg/L}} Q_{\text{MGD}}) \\ \times \left(8.345 \frac{\text{lbm-L}}{\text{MG-mg}}\right) \\ -1.42 P_{x,\text{lbm/day}} \end{array} \right) \quad 30.51(b)$$

$$q = (\text{LHV}) V_{\text{fuel}} \qquad 30.52$$

20. HEAT TRANSFER AND LOSS

$$q = m_{\text{sludge}} c_p (T_2 - T_1)$$
$$= V_{\text{sludge}} \rho_{\text{sludge}} c_p (T_2 - T_1) \quad \begin{bmatrix} \text{sensible} \\ \text{energy} \end{bmatrix} \quad 30.53$$

$$q = h A_{\text{surface}} (T_{\text{surface}} - T_{\text{air}}) \quad \begin{bmatrix} \text{convective} \\ \text{energy} \end{bmatrix} \quad 30.54$$

$$q = U A_{\text{surface}} (T_{\text{contents}} - T_{\text{air}}) \quad \begin{bmatrix} \text{total} \\ \text{energy} \end{bmatrix} \quad 30.55$$

$$q = \frac{k A (T_{\text{inner}} - T_{\text{outer}})}{L} \quad \begin{bmatrix} \text{conductive} \\ \text{energy} \end{bmatrix} \quad 30.56$$

21. SLUDGE DEWATERING

$$V_{\text{press}} \rho_{\text{filter cake}} s_{\text{filter cake}}$$
$$= V_{\text{sludge,per cycle}} s_{\text{sludge}} \rho_{\text{sludge}}$$
$$= V_{\text{sludge,per cycle}} s_{\text{sludge}} \rho_{\text{water}} \text{SG}_{\text{sludge}} \quad 30.57$$

$$G = \frac{\omega^2 r}{g} \quad [\text{centrifuge}] \qquad 30.58$$

CERM Chapter 31
Municipal Solid Waste

Chapter, section, equation, figure, and table numbers correspond to CERM. For additional study material, go to the corresponding chapter and section number in CERM.

3. LANDFILL CAPACITY

$$V_c = (\text{CF}) V_o \qquad 31.1$$

$$\Delta V_{\text{day}} = \frac{NG(\text{LF})}{\gamma} = \frac{NG(\text{LF}) g_c}{\rho g} \qquad 31.2(b)$$

$$\text{LF} = \frac{V_{\text{MSW}} + V_{\text{cover soil}}}{V_{\text{MSW}}} \qquad 31.3$$

11. LANDFILL GAS

$$p_t = p_{\text{CH}_4} + p_{\text{CO}_2} + p_{\text{H}_2\text{O}} + p_{\text{N}_2} + p_{\text{other}} \qquad 31.8$$

$$p_i = B_i p_t \qquad 31.9$$

13. LEACHATE MIGRATION FROM LANDFILLS

$$Q = KiA \qquad 31.13$$

$$i = \frac{dH}{dL} \qquad 31.14$$

19. INCINERATION OF MUNICIPAL SOLID WASTE

$$\text{HRR} = \frac{(\text{fueling rate})(\text{HV})}{\text{total effective grate area}} \qquad 31.24$$

CERM Chapter 32
Pollutants in the Environment

Chapter, section, equation, figure, and table numbers correspond to CERM. For additional study material, go to the corresponding chapter and section number in CERM.

38. SMOKE

$$\text{optical density} = \log_{10} \left(\frac{1}{1 - \text{opacity}} \right) \qquad 32.1$$

CERM Chapter 34
Environmental Remediation

Chapter, section, equation, figure, and table numbers correspond to CERM. For additional study material, go to the corresponding chapter and section number in CERM.

16. CYCLONE SEPARATORS

$$S = \frac{\text{v}_{\text{inlet}}^2}{rg} \qquad 34.2$$

$$h = \frac{KBH\text{v}_{\text{inlet}}^2}{2gD^2} \qquad 34.3$$

18. ELECTROSTATIC PRECIPITATORS

$$\eta = 1 - K^{Ct} = 1 - e^{\frac{-wA_p}{Q}} \qquad 34.4$$

47. STRIPPING, AIR

$$p_{\text{A}} = H_{\text{A}} x_{\text{A}} \qquad 34.6$$

$$R = \frac{HG}{L} \qquad 34.7$$

Geotechnical

CERM Chapter 35
Soil Properties and Testing

Chapter, section, equation, figure, and table numbers correspond to CERM. For additional study material, go to the corresponding chapter and section number in CERM.

1. SOIL PARTICLE SIZE DISTRIBUTION

$$C_u = \frac{D_{60}}{D_{10}} \qquad 35.1$$

$$C_z = \frac{(D_{30})^2}{D_{10}D_{60}} \qquad 35.2$$

3. AASHTO SOIL CLASSIFICATION

$$I_g = (F_{200} - 35)\big(0.2 + 0.005(\text{LL} - 40)\big) \\ + 0.01(F_{200} - 15)(\text{PI} - 10) \qquad 35.3$$

5. MASS-VOLUME RELATIONSHIPS

Table 35.7 *Soil Indexing Formulas*

property		saturated sample (m_s, m_w, SG are known)	unsaturated sample (m_s, m_w, SG, V_t are known)	supplementary formulas relating measured and computed factors			
volume components							
V_s	volume of solids		$\dfrac{m_s}{(\text{SG})\rho_w}$	$V_t - (V_g + V_w)$	$V_t(1-n)$	$\dfrac{V_t}{1+e}$	$\dfrac{V_v}{e}$
V_w	volume of water		$\dfrac{m_w}{\rho_w^*}$	$V_v - V_g$	SV_v	$\dfrac{SVe}{1+e}$	$SV_s e$
V_g	volume of gas or air	zero	$V_t - (V_s + V_w)$	$V_v - V_w$	$(1-S)V_v$	$\dfrac{(1-S)Ve}{1+e}$	$(1-S)V_s e$
V_v	volume of voids	$\dfrac{m_w}{\rho_w^*}$	$V_t - \dfrac{m_s}{(\text{SG})\rho_w}$	$V_t - V_s$	$\dfrac{V_s n}{1-n}$	$\dfrac{Ve}{1+e}$	$V_s e$
V_t	total volume of sample	$V_s + V_w$	measured $(V_g + V_w + V_s)$	$V_s + V_g + V_w$	$\dfrac{V_s}{1-n}$	$V_s(1+e)$	$\dfrac{V_v(1+e)}{e}$
n	porosity		$\dfrac{V_v}{V_t}$	$1 - \dfrac{V_s}{V_t}$	$1 - \dfrac{m_s}{(\text{SG})V_t\rho_w}$	$\dfrac{e}{1+e}$	
e	void ratio		$\dfrac{V_v}{V_s}$	$\dfrac{V_t}{V_s} - 1$	$\dfrac{(\text{SG})V_t\rho_w}{m_s} - 1$	$\dfrac{m_w(\text{SG})}{m_s S}$	$\dfrac{n}{1-n}\bigg\|w\Big(\dfrac{\text{SG}}{\text{S}}\Big)$

(continued)

Table 35.7 *Soil Indexing Formulas (continued)*

property	saturated sample (m_s, m_w, SG are known)	unsaturated sample (m_s, m_w, SG, V_t are known)	supplementary formulas relating measured and computed factors			
mass for specific sample						
m_s mass of solids		measured	$\dfrac{m_t}{1+w}$	$(SG)V_t\rho_w(1-n)$	$\dfrac{m_w(SG)}{eS}$	$V_s(SG)\rho_w$
m_w mass of water		measured	wm_s	$S\rho_w V_v$	$\dfrac{em_s S}{SG}$	$V_t\rho_d w$
m_t total mass of sample		m_s+m_w	$m_s(1+w)$			
mass for sample of unit volume (density)						
ρ_d dry density	$\dfrac{m_s}{V_s+V_w}$	$\dfrac{m_s}{V_g+V_w+V_s}$	$\dfrac{m_t}{V_t(1+w)}$	$\dfrac{(SG)\rho_w}{1+e}$	$\dfrac{(SG)\rho_w}{1+\dfrac{w(SG)}{S}}$	$\dfrac{\rho}{1+w}$
ρ wet density	$\dfrac{m_s+m_w}{V_s+V_w}$	$\dfrac{m_s+m_w}{V_t}$	$\dfrac{m_t}{V_t}$	$\dfrac{(SG+Se)\rho_w}{1+e}$	$\dfrac{(1+w)\rho_w}{\dfrac{w}{S}+\dfrac{1}{SG}}$	$\rho_d(1+w)$
ρ_{sat} saturated density	$\dfrac{m_s+m_w}{V_s+V_w}$	$\dfrac{m_s+V_v\rho_w}{V_t}$	$\dfrac{m_s}{V_t}+\left(\dfrac{e}{1+e}\right)\rho_w$	$\dfrac{(SG+e)\rho_w}{1+e}$	$\dfrac{(1+w)\rho_w}{w+\dfrac{1}{SG}}$	
ρ_b buoyant (submerged) density		$\rho_{sat}-\rho_w^*$	$\dfrac{m_s}{V_t}-\left(\dfrac{1}{1+e}\right)\rho_w^*$	$\left(\dfrac{SG+e}{1+e}-1\right)\rho_w^*$	$\left(\dfrac{1-\dfrac{1}{SG}}{w+\dfrac{1}{SG}}\right)\rho_w^*$	
combined relations						
w water content		$\dfrac{m_w}{m_s}$	$\dfrac{m_t}{m_s}-1$	$\dfrac{Se}{SG}$	$S\left(\dfrac{\rho_w^*}{\rho_d}-\dfrac{1}{SG}\right)$	
S degree of saturation	100%	$\dfrac{V_w}{V_v}$	$\dfrac{m_w}{V_v\rho_w^*}$	$\dfrac{w(SG)}{e}$	$\dfrac{w}{\dfrac{\rho_w^*}{\rho_d}-\dfrac{1}{SG}}$	
SG specific gravity of solids		$\dfrac{m_s}{V_s\rho_w}$	$\dfrac{Se}{w}$			

ρ_w is the density of water. Where noted with an asterisk (*), use the actual density of water at the recorded temperature. In other cases, use 62.4 lbm/ft^3 or 1000 kg/m^3.

7. EFFECTIVE STRESS

$$\sigma' = \sigma - u \qquad 35.17$$

10. CONE PENETROMETER TEST

$$f_R = \frac{q_s}{q_c} \times 100\% \qquad 35.18$$

11. PROCTOR TEST

$$\text{RC} = \frac{\rho_d}{\rho_d^*} \times 100\% \qquad 35.19$$

12. MODIFIED PROCTOR TEST

$$\rho_z = \frac{m_s}{V_w + V_s} \qquad 35.20$$

$$\rho_z = \frac{\rho_w}{w + \dfrac{1}{\text{SG}}} \qquad 35.21$$

$$\rho_s = (\text{SG})\rho_w \qquad 35.22$$

14. ATTERBERG LIMIT TESTS

$$\text{PI} = \text{LL} - \text{PL} \qquad 35.23$$

$$\text{SI} = \text{PL} - \text{SL} \qquad 35.24$$

$$\text{LI} = \frac{w - \text{PL}}{\text{PI}} \qquad 35.25$$

15. PERMEABILITY TESTS

$$Q = \text{v}A_{\text{gross}} \qquad 35.26$$

$$\text{v} = Ki \qquad 35.27$$

$$K_{\text{cm/s}} \approx C(D_{10,\text{mm}})^2 \qquad 35.28$$

$$K = \frac{VL}{hAt} \qquad \text{[constant head]} \quad 35.29$$

$$K = \left(\frac{A'L}{At}\right) \ln\left(\frac{h_i}{h_f}\right) \qquad \text{[falling head]} \quad 35.30$$

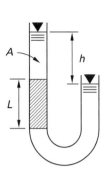 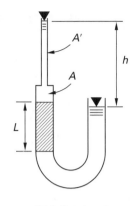

(a) constant head   (b) falling head

Figure 35.8 *Permeameters*

$$\text{F} = \frac{11r}{2} \qquad \text{[auger hole]} \quad 35.31$$

$$K = \left(\frac{\pi r^2}{Ft}\right) \ln\left(\frac{h_i}{h_f}\right)$$

$$= \left(\frac{2\pi r}{11t}\right) \ln\left(\frac{h_i}{h_f}\right) \qquad \text{[auger hole]} \quad 35.32$$

16. CONSOLIDATION TESTS

$$\text{OCR} = \frac{p'_{\text{max}}}{p'_o} \qquad 35.33$$

$$C_c = -\frac{e_1 - e_2}{\log_{10}\left(\dfrac{p_1}{p_2}\right)} \qquad 35.34$$

$$C_{\epsilon c} = \frac{C_c}{1 + e_0} \qquad 35.35$$

$$C_c \approx (0.009)(\text{LL} - 10) \qquad 35.36$$

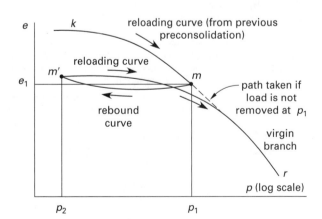

Figure 35.10 *e-log p Curve*

17. DIRECT SHEAR TEST

$$S = \tau = c + \sigma \tan\phi \qquad 35.37$$

18. TRIAXIAL STRESS TEST

$$\sigma_\theta = \tfrac{1}{2}(\sigma_A + \sigma_R)$$
$$+ \tfrac{1}{2}(\sigma_A - \sigma_R)\cos 2\theta \qquad 35.38$$

$$\tau_\theta = \tfrac{1}{2}(\sigma_A - \sigma_R)\sin 2\theta \qquad 35.39$$

$$\frac{\sigma_1}{\sigma_3} = \frac{1 + \sin\phi}{1 - \sin\phi} \quad [c = 0] \qquad 35.40$$

$$\alpha = 45° + \tfrac{1}{2}\phi \qquad 35.41$$

$$S_u = c = \frac{\sigma_D}{2} \qquad 35.42$$

$$s = c' + \sigma'\tan\phi' \qquad 35.43$$

20. UNCONFINED COMPRESSIVE STRENGTH TEST

$$S_{uc} = \frac{P}{A} \qquad 35.44$$

$$s_u = \frac{S_{uc}}{2} \qquad 35.45$$

21. SENSITIVITY

$$S_t = \frac{S_{undisturbed}}{S_{remolded}} \qquad 35.46$$

22. CALIFORNIA BEARING RATIO TEST

$$\text{CBR} = \frac{\text{actual load}}{\text{standard load}} \times 100 \qquad 35.47$$

CERM Chapter 36
Shallow Foundations

Chapter, section, equation, figure, and table numbers correspond to CERM. For additional study material, go to the corresponding chapter and section number in CERM.

5. GENERAL BEARING CAPACITY EQUATION

$$q_{ult} = \tfrac{1}{2}\gamma B N_\gamma + c N_c + (p_q + \gamma D_f)N_q \qquad 36.1(b)$$

Table 36.2 Terzaghi Bearing Capacity Factors for General Shear[a]

ϕ	N_c	N_q	N_γ
0	5.7	1.0	0.0
5	7.3	1.6	0.5
10	9.6	2.7	1.2
15	12.9	4.4	2.5
20	17.7	7.4	5.0
25	25.1	12.7	9.7
30	37.2	22.5	19.7
34	52.6	36.5	35.0
35	57.8	41.4	42.4
40	95.7	81.3	100.4
45	172.3	173.3	297.5
48	258.3	287.9	780.1
50	347.5	415.1	1153.2

[a]Curvilinear interpolation may be used. Do not use linear interpolation.

Table 36.3 Meyerhof and Vesic Bearing Capacity Factors for General Shear[a]

ϕ	N_c	N_q	N_γ	$N_\gamma{}^b$
0	5.14	1.0	0.0	0.0
5	6.5	1.6	0.07	0.5
10	8.3	2.5	0.37	1.2
15	11.0	3.9	1.1	2.6
20	14.8	6.4	2.9	5.4
25	20.7	10.7	6.8	10.8
30	30.1	18.4	15.7	22.4
32	35.5	23.2	22.0	30.2
34	42.2	29.4	31.2	41.1
36	50.6	37.7	44.4	56.3
38	61.4	48.9	64.1	78.0
40	75.3	64.2	93.7	109.4
42	93.7	85.4	139.3	155.6
44	118.4	115.3	211.4	224.6
46	152.1	158.5	328.7	330.4
48	199.3	222.3	526.5	496.0
50	266.9	319.1	873.9	762.9

[a]Curvilinear interpolation may be used. Do not use linear interpolation.
[b]As predicted by the Vesic equation, $N_\gamma = 2(N_q + 1)\tan\phi$.

Table 36.4 N_c Bearing Capacity Factor Multipliers for Various Values of B/L

B/L	multiplier
1 (square)	1.25
0.5	1.12
0.2	1.05
0.0	1.00
1 (circular)	1.20

Table 36.5 N_γ Multipliers for Various Values of B/L

B/L	multiplier
1 (square)	0.85
0.5	0.90
0.2	0.95
0.0	1.00
1 (circular)	0.70

$$q_{\text{net}} = q_{\text{ult}} - \gamma D_f \qquad \text{36.3(b)}$$

$$q_a = \frac{q_{\text{net}}}{F} \qquad \text{36.4}$$

6. BEARING CAPACITY OF CLAY

$$S_u = c = \frac{S_{\text{uc}}}{2} \qquad \text{36.5}$$

$$q_{\text{ult}} = cN_c + \gamma D_f \qquad \text{36.6(b)}$$

$$q_{\text{net}} = q_{\text{ult}} - \gamma D_f = cN_c \qquad \text{36.7(b)}$$

$$q_a = \frac{q_{\text{net}}}{F} \qquad \text{36.8}$$

7. BEARING CAPACITY OF SAND

$$q_{\text{ult}} = \tfrac{1}{2}B\gamma N_\gamma + (p_q + \gamma D_f)N_q \qquad \text{36.9(b)}$$

$$q_{\text{net}} = q_{\text{ult}} - \gamma D_f$$
$$= \tfrac{1}{2}B\gamma N_\gamma + \gamma D_f(N_q - 1) \qquad \text{36.10(b)}$$

$$q_a = \frac{q_{\text{net}}}{F}$$
$$= \left(\frac{B}{F}\right)\left(\tfrac{1}{2}\gamma N_\gamma + \gamma(N_q - 1)\left(\frac{D_f}{B}\right)\right) \qquad \text{36.11(b)}$$

$$q_a = 0.11C_nN \quad [\text{in tons/ft}^2] \qquad \text{36.12}$$

9. EFFECTS OF WATER TABLE ON FOOTING DESIGN

$$q_{\text{ult}} = \tfrac{1}{2}\gamma_b BN_\gamma + \gamma_d D_f N_q \quad \begin{bmatrix}\text{sand; water table at}\\ \text{base of footing; } c = 0\end{bmatrix}$$
$$\text{36.13(b)}$$

$$q_{\text{ult}} = \tfrac{1}{2}\gamma_b BN_\gamma + \gamma_b D_f N_q \quad \begin{bmatrix}\text{sand; water table}\\ \text{at surface; } c = 0\end{bmatrix}$$
$$\text{36.14(b)}$$

$$\left(p_q + \gamma_d D_w + \left(\gamma_d - 62.4\,\frac{\text{lbf}}{\text{ft}^3}\right)(D_f - D_w)\right)N_q$$
$$= \left(p_q + \gamma D_f + \left(62.4\,\frac{\text{lbf}}{\text{ft}^3}\right)(D_w - D_f)\right)N_q$$
$$\begin{bmatrix}\text{sand; water table between}\\ \text{base and surface; } c = 0\end{bmatrix} \text{36.15(b)}$$

$$q_{\text{ult}} = \tfrac{1}{2}\gamma_d BN_\gamma + \gamma_d DN_q$$
$$\begin{bmatrix}\text{sand; } D_w > D_f + B;\\ c = 0\end{bmatrix} \text{36.16(b)}$$

10. ECCENTRIC LOADS ON RECTANGULAR FOOTINGS

$$\epsilon_B = \frac{M_B}{P}; \quad \epsilon_L = \frac{M_L}{P} \qquad \text{36.17}$$

$$L' = L - 2\epsilon_L; \quad B' = B - 2\epsilon_B \qquad \text{36.18}$$

$$A' = L'B' \qquad \text{36.19}$$

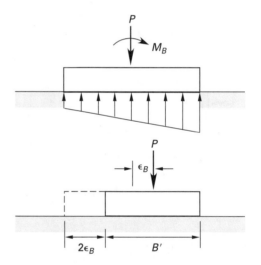

Figure 36.6 Footing with Overturning Moment

$$p_{\text{max}}, p_{\text{min}} = \left(\frac{P}{BL}\right)\left(1 \pm \frac{6\epsilon}{B}\right) \qquad \text{36.20}$$

12. RAFTS ON CLAY

$$F = \frac{cN_c}{\dfrac{\text{total load}}{\text{raft area}} - \gamma D_f} \qquad \text{36.21(b)}$$

13. RAFTS ON SAND

$$q_a = 0.22C_nN \quad [\text{in tons/ft}^2] \qquad \text{36.22}$$

$$p = \frac{\text{total load}}{\text{raft area}} - \gamma D_f \qquad \text{36.23(b)}$$

CERM Chapter 37
Rigid Retaining Walls

Chapter, section, equation, figure, and table numbers correspond to CERM. For additional study material, go to the corresponding chapter and section number in CERM.

3. EARTH PRESSURE

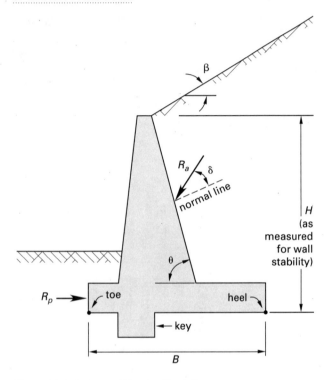

Figure 37.2 *Active and Passive Earth Pressure*

$$\alpha = 45° + \frac{\phi}{2} \quad \text{[Rankine]} \qquad 37.1$$

$$\alpha_a = 90° - \phi - \arctan\left(-\tan\phi \right.$$
$$\left. + \frac{\sqrt{\tan\phi(\tan\phi + \cot\phi)(1 + \tan\delta\cot\phi)}}{1 + \tan\delta(\tan\phi + \cot\phi)} \right)$$
$$\text{[Coulomb]} \quad 37.2$$

4. VERTICAL SOIL PRESSURE

$$p_v = \gamma H \qquad 37.3(b)$$

5. ACTIVE EARTH PRESSURE

$$p_a = p_v k_a - 2c\sqrt{k_a} \qquad 37.4$$

$$k_a = \frac{\sin^2(\theta + \phi)}{\sin^2\theta \sin(\theta - \delta)\left(1 + \sqrt{\dfrac{\sin(\phi + \delta)\sin(\phi - \beta)}{\sin(\theta - \delta)\sin(\theta + \beta)}}\right)^2}$$
$$\text{[Coulomb]} \quad 37.5$$

$$k_a = \cos\beta\left(\frac{\cos\beta - \sqrt{\cos^2\beta - \cos^2\phi}}{\cos\beta + \sqrt{\cos^2\beta - \cos^2\phi}}\right) \quad \text{[Rankine]}$$
$$37.6$$

$$k_a = \frac{1}{k_p} = \tan^2\left(45° - \frac{\phi}{2}\right)$$
$$= \frac{1 - \sin\phi}{1 + \sin\phi} \quad \begin{bmatrix} \text{Rankine: horizontal} \\ \text{backfill; vertical face} \end{bmatrix} \quad 37.7$$

$$p_a = p_v - 2c \quad [\phi = 0] \qquad 37.8$$

$$p_a = k_a p_v \quad [c = 0] \qquad 37.9$$

$$R_a = \tfrac{1}{2}p_a H = \tfrac{1}{2}k_a\gamma H^2 \qquad 37.10(b)$$

$$\theta_R = 90° \text{ from the wall} \quad \text{[Rankine]} \qquad 37.11$$

$$\theta_R = 90° - \delta \text{ from the wall} \quad \text{[Coulomb]} \qquad 37.12$$

6. PASSIVE EARTH PRESSURE

$$p_p = p_v k_p + 2c\sqrt{k_p} \qquad 37.13$$

$$k_p = \frac{\sin^2(\theta - \phi)}{\sin^2\theta \sin(\theta + \delta)\left(1 - \sqrt{\dfrac{\sin(\phi + \delta)\sin(\phi + \beta)}{\sin(\theta + \delta)\sin(\theta + \beta)}}\right)^2}$$
$$\text{[Coulomb]} \quad 37.14$$

$$k_p = \cos\beta\left(\frac{\cos\beta + \sqrt{\cos^2\beta - \cos^2\phi}}{\cos\beta - \sqrt{\cos^2\beta - \cos^2\phi}}\right) \quad \text{[Rankine]}$$
$$37.15$$

$$k_p = \frac{1}{k_a} = \tan^2\left(45° + \frac{\phi}{2}\right)$$
$$= \frac{1 + \sin\phi}{1 - \sin\phi} \quad \begin{bmatrix} \text{Rankine: horizontal} \\ \text{backfill; vertical face} \end{bmatrix} \quad 37.16$$

$$p_p = p_v + 2c \quad [\phi = 0] \qquad 37.17$$

$$p_p = k_p p_v \quad [c = 0] \qquad 37.18$$

$$R_p = \tfrac{1}{2}p_p H = \tfrac{1}{2}k_p\gamma H^2 \qquad 37.19(b)$$

7. AT-REST SOIL PRESSURE

$$p_o = k_o p_v \qquad 37.20$$

$$k_o \approx 1 - \sin \phi \qquad 37.21$$

$$R_o = \tfrac{1}{2} k_o \gamma H^2 \qquad 37.22(b)$$

9. SURCHARGE LOADING

$$p_q = k_a q \qquad 37.27$$

$$R_q = k_a q H \times (\text{wall width}) \qquad 37.28$$

$$p_q = \frac{1.77 V_q m^2 n^2}{H^2 (m^2 + n^2)^3} \quad [m > 0.4] \qquad 37.29$$

$$p_q = \frac{0.28 V_q n^2}{H^2 (0.16 + n^2)^3} \quad [m \leq 0.4] \qquad 37.30$$

$$R_q \approx \frac{0.78 V_q}{H} \quad [m = 0.4] \qquad 37.31$$

$$R_q \approx \frac{0.60 V_q}{H} \quad [m = 0.5] \qquad 37.32$$

$$R_q \approx \frac{0.46 V_q}{H} \quad [m = 0.6] \qquad 37.33$$

$$m = \frac{x}{H} \qquad 37.34$$

$$n = \frac{y}{H} \qquad 37.35$$

$$p_q = \frac{4 L_q m^2 n}{\pi H (m^2 + n^2)^2} \quad [m > 0.4] \qquad 37.36$$

$$R_q = \frac{0.64 L_q}{m^2 + 1} \qquad 37.37$$

$$p_q = \frac{0.203 L_q n}{H (0.16 + n^2)^2} \quad [m \leq 0.4] \qquad 37.38$$

$$R_q = 0.55 L_q \qquad 37.39$$

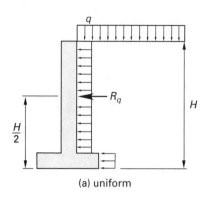

(a) uniform

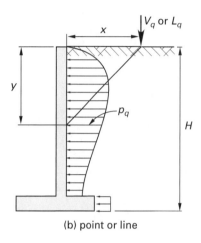

(b) point or line

Figure 37.3 *Surcharges*

10. EFFECTIVE STRESS

$$\mu = \gamma_w h \qquad 37.40(b)$$

$$\gamma_{\text{sat}} = \gamma_{\text{dry}} + n \gamma_w$$

$$= \gamma_{\text{dry}} + \left(\frac{e}{1 + e} \right) \gamma_w \qquad 37.41(b)$$

$$p_v = \gamma_{\text{sat}} H - \gamma_w h \qquad 37.42(b)$$

$$p_h = \gamma_w h + k_a (\gamma_{\text{sat}} H - \gamma_w h)$$

$$= k_a \gamma_{\text{sat}} H + (1 - k_a) \gamma_w h \qquad 37.43(b)$$

$$\gamma_{\text{eq}} = (1 - k_a) \gamma_w \qquad 37.44(b)$$

11. CANTILEVER RETAINING WALLS: ANALYSIS

$$W_i = \gamma_i A_i \qquad 37.45(b)$$

$$M_{\text{toe}} = \sum W_i x_i - R_{a,h} y_a + R_{a,v} x_a \qquad 37.46$$

$$x_R = \frac{M_{\text{toe}}}{\sum (W_i + R_{a,v})} \qquad 37.47$$

$$\epsilon = \left| \frac{B}{2} - x_R \right| \qquad 37.48$$

$$F_{OT} = \frac{M_{resisting}}{M_{overturning}}$$

$$= \frac{\sum W_i x_i + R_{a,v} x_{a,v}}{R_{a,h} y_{a,h}} \qquad 37.49$$

$$p_{v,max}, \ p_{v,min} = \left(\frac{\sum W_i + R_{a,v}}{B} \right)$$

$$\times \left(1 \pm \left(\frac{6\epsilon}{B} \right) \right) \qquad 37.50$$

$$R_{SL} = (\Sigma W_i + R_{a,v}) \tan \delta + c_A B \qquad 37.52$$

$$F_{SL} = \frac{R_{SL}}{R_{a,h}} \qquad 37.53$$

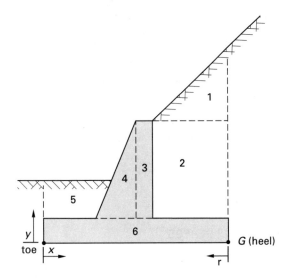

^{a}Some retaining walls may not have all elements.

Figure 37.4 *Elements Contributing to Vertical Force^a (step 3)*

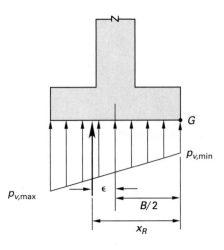

Figure 37.5 *Resultant Distribution on the Base (step 7)*

12. RETAINING WALLS: DESIGN

$$R_{a,h} \left(\frac{H}{3} \right) \approx (\text{soil weight}) \left(\frac{L}{2} \right) \qquad 37.54$$

$$B \approx \tfrac{3}{2} L \qquad 37.55$$

$$8 < \frac{H}{t_{stem}} < 12 \qquad 37.56$$

$$10 < \frac{H}{t_{base}} < 14 \qquad 37.57$$

CERM Chapter 38
Piles and Deep Foundations

Chapter, section, equation, figure, and table numbers correspond to CERM. For additional study material, go to the corresponding chapter and section number in CERM.

1. INTRODUCTION

$$Q_{ult} = Q_p + Q_f \qquad 38.1$$

$$Q_a = \frac{Q_{ult}}{F} \qquad 38.2$$

2. PILE CAPACITY FROM DRIVING DATA

$$Q_{a,lbf} = \frac{2W_{hammer,lbf} H_{fall,ft}}{S_{in} + 1} \quad [\text{drop hammer}] \quad 38.3$$

$$Q_{a,lbf} = \frac{2W_{hammer,lbf} H_{fall,ft}}{S_{in} + 0.1}$$

$$\left[\begin{array}{l} \text{single-acting steam hammer;} \\ \text{driven weight} < \text{striking weight} \end{array} \right] \quad 38.4$$

$$Q_{a,lbf} = \frac{2W_{hammer,lbf} H_{fall,ft}}{S_{in} + 0.1 \left(\dfrac{W_{driven}}{W_{hammer}} \right)}$$

$$\left[\begin{array}{l} \text{single-acting steam hammer;} \\ \text{driven weight} > \text{striking weight} \end{array} \right] \quad 38.5$$

$$Q_{a,lbf} = \frac{2E_{ft-lbf}}{S_{in} + 0.1}$$

$$\left[\begin{array}{l} \text{double-acting steam hammer;} \\ \text{driven weight} < \text{striking weight} \end{array} \right] \quad 38.6$$

$$Q_{a,lbf} = \frac{2E_{ft-lbf}}{S_{in} + 0.1 \left(\dfrac{W_{driven}}{W_{hammer}} \right)}$$

$$\left[\begin{array}{l} \text{double-acting steam hammer;} \\ \text{driven weight} > \text{striking weight} \end{array} \right] \quad 38.7$$

3. THEORETICAL POINT-BEARING CAPACITY

$$Q_p = A_p \left(\tfrac{1}{2} \gamma B N_\gamma + c N_c + \gamma D_f N_q \right) \qquad 38.8(b)$$

$$Q_p = A_p \gamma D N_q \quad [\text{cohesionless}; \ D \le D_c] \qquad 38.9(b)$$

$$Q_p = A_p c N_c \approx 9 A_p c \quad [\text{cohesive}] \qquad 38.10$$

4. THEORETICAL SKIN-FRICTION CAPACITY

$$Q_f = A_s f_s = p f_s L_e$$
$$= p f_s (L - \text{seasonal variation}) \qquad 38.11$$

$$Q_f = p \sum f_{s,i} L_{e,i} \qquad 38.12$$

$$f_s = c_A + \sigma_h \tan \delta \qquad 38.13$$

$$f_s = \alpha c \qquad 38.14$$

$$\sigma_h = k_s \sigma'_v = k_s (\gamma D - \mu) \qquad 38.15(b)$$

$$\mu = \gamma_w h \qquad 38.16(b)$$

$$Q_f = p k_s \tan \delta \sum L_i \sigma' \quad [\text{cohesionless}] \qquad 38.17$$

$$Q_f = p \sum c_A L_i \quad [\text{cohesive}] \qquad 38.18$$

$$Q_f = p \beta \sigma' L \quad [\text{cohesive}] \qquad 38.19$$

7. CAPACITY OF PILE GROUPS

$$Q_s = 2(b + w) L_e c_1 \qquad 38.20$$

$$Q_p = 9 c_2 b w \qquad 38.21$$

$$Q_{\text{ult}} = Q_s + Q_p \qquad 38.22$$

$$Q_a = \frac{Q_{\text{ult}}}{F} \qquad 38.23$$

$$\eta_G = \frac{\text{group capacity}}{\sum \text{individual capacities}} \qquad 38.24$$

CERM Chapter 39
Temporary Excavations

Chapter, section, equation, figure, and table numbers correspond to CERM. For additional study material, go to the corresponding chapter and section number in CERM.

2. BRACED CUTS IN SAND

$$p_{\text{max}} = 0.65 k_a \gamma H \qquad 39.1(b)$$

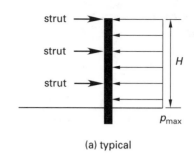

(a) typical

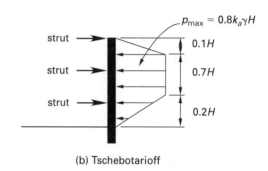

(b) Tschebotarioff

Figure 39.2 *Cuts in Sand*

3. BRACED CUTS IN STIFF CLAY

$$0.2 \gamma H \le p_{\text{max}} \le 0.4 \gamma H \qquad 39.2(b)$$

$$p_{\text{max}} = k_a \gamma H \qquad 39.3(b)$$

$$k_a = 1 - \frac{4c}{\gamma H} \qquad 39.4(b)$$

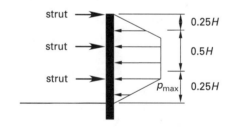

Figure 39.3 *Cuts in Stiff Clay*

4. BRACED CUTS IN SOFT CLAY

$$p_{\text{max}} = \gamma H - 4c \qquad 39.5(b)$$

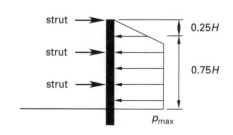

Figure 39.4 *Cuts in Soft Clay*

6. ANALYSIS/DESIGN OF BRACED EXCAVATIONS

$$S = \frac{M_{\text{max,sheet piling}}}{F_b} \qquad 39.6$$

7. STABILITY OF BRACED EXCAVATIONS IN CLAY

$$H_c = \frac{5.7c}{\gamma - \sqrt{2}\left(\dfrac{c}{B}\right)} \quad [H < B] \qquad 39.7(b)$$

$$H_c = \frac{N_c c}{\gamma} \quad [H > B] \qquad 39.8(b)$$

$$F = \frac{N_c c}{\gamma H + q} \qquad 39.9(b)$$

8. STABILITY OF BRACED EXCAVATIONS IN SAND

$$F = 2N_\gamma k_a \tan \phi \qquad 39.10$$

$$F = 2N_\gamma \left(\frac{\gamma_{\text{submerged}}}{\gamma_{\text{drained}}}\right) k_a \tan \phi \qquad 39.11(b)$$

10. ANALYSIS/DESIGN OF FLEXIBLE BULKHEADS

$$y = \frac{k_p D^2 - k_a (H + D)^2}{(k_p - k_a)(H + 2D)} \qquad 39.12$$

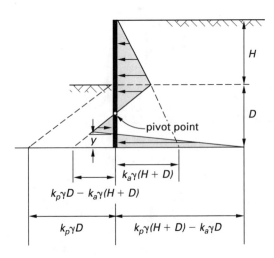

Figure 39.5 *Cantilever Wall in Uniform Granular Soil*

12. ANALYSIS/DESIGN OF TIED BULKHEADS

$$f = \frac{M}{S} \qquad 39.13$$

$$S = \frac{M}{F_b} \qquad 39.14$$

CERM Chapter 40
Special Soil Topics

> Chapter, section, equation, figure, and table numbers correspond to CERM. For additional study material, go to the corresponding chapter and section number in CERM.

1. PRESSURE FROM APPLIED LOADS: BOUSSINESQ'S EQUATION

$$\Delta p_v = \frac{3h^3 P}{2\pi z^5}$$

$$= \left(\frac{3P}{2\pi h^2}\right)\left(\frac{1}{1 + \left(\dfrac{r}{h}\right)^2}\right)^{\frac{5}{2}} \quad [h > 2B] \qquad 40.1$$

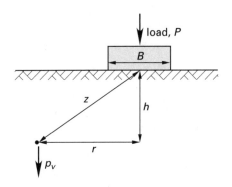

Figure 40.1 *Pressure at a Point*

2. PRESSURE FROM APPLIED LOADS: ZONE OF INFLUENCE

$$A = \left(B + 2(h \cot 60°)\right)\left(L + 2(h \cot 60°)\right) \qquad 40.2$$

$$A = (B + h)(L + h) \qquad 40.3$$

$$\Delta p_v = \frac{P}{A} \qquad 40.4$$

3. PRESSURE FROM APPLIED LOADS: INFLUENCE CHART

$$\Delta p_v = \text{(influence value)(no. of squares)}$$
$$\times \text{(applied pressure)} \qquad 40.5$$

5. CLAY CONDITION

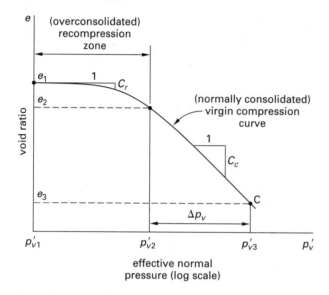

Figure 40.4 *Consolidation Curve for Clay*

6. CONSOLIDATION PARAMETERS

$$C_r = \frac{-(e_1 - e_2)}{\log_{10}\left(\dfrac{p_1}{p_2}\right)} \qquad 40.6$$

$$C_c = \frac{-(e_2 - e_3)}{\log_{10}\left(\dfrac{p_2}{p_3}\right)} = \frac{\Delta e}{\log_{10}\left(\dfrac{p_2}{p_2 + \Delta p_v}\right)} \qquad 40.7$$

$$C_c \approx 1.15(e_o - 0.35) \quad \text{[clays]} \qquad 40.8$$

$$C_c \approx 0.009(\text{LL} - 10) \quad \begin{bmatrix} \text{overconsolidated} \\ \text{clays} \end{bmatrix} \qquad 40.9$$

$$C_c \approx 0.0155w \quad \text{[organic soils]} \qquad 40.10$$

$$C_c \approx (1 + e_o)\big(0.1 + 0.006(w_n - 25)\big) \quad \begin{bmatrix} \text{varved} \\ \text{clays} \end{bmatrix} \qquad 40.11$$

$$\text{CR} = \frac{C_c}{1 + e_o} \qquad 40.12$$

$$\text{RR} = \frac{C_r}{1 + e_o} \qquad 40.13$$

$$e = w(\text{SG}) \quad \text{[saturated]} \qquad 40.14$$

7. PRIMARY CONSOLIDATION

$$S_{\text{primary}} = \frac{H\Delta e}{1 + e_o}$$

$$= \frac{HC_r \log_{10}\left(\dfrac{p_o' + \Delta p_v'}{p_o'}\right)}{1 + e_o}$$

$$= H(\text{RR}) \log_{10}\left(\frac{p_o' + \Delta p_v'}{p_o'}\right) \quad \text{[overconsolidated]}$$
$$40.15$$

$$S_{\text{primary}} = \frac{H\Delta e}{1 + e_o}$$

$$= \frac{HC_c \log_{10}\left(\dfrac{p_o' + \Delta p_v'}{p_o'}\right)}{1 + e_o}$$

$$= H(\text{CR}) \log_{10}\left(\frac{p_o' + \Delta p_v'}{p_o'}\right) \quad \begin{bmatrix} \text{normally} \\ \text{consolidated} \end{bmatrix}$$
$$40.16$$

$$p_v = \gamma_{\text{layer}}H \quad \text{[above GWT]} \qquad 40.17$$
$$p_v' = \gamma_{\text{layer}}H - \gamma_{\text{water}}h$$
$$\quad = p_v - \mu \quad \text{[below GWT]} \qquad 40.18$$
$$\mu = \gamma_{\text{water}}h \qquad 40.19$$

8. PRIMARY CONSOLIDATION RATE

$$t = \frac{T_v H_d^2}{C_v} \qquad 40.20$$

$$C_v = \frac{K(1 + e_o)}{a_v \gamma_{\text{water}}} \qquad 40.21$$

$$a_v = \frac{-(e_2 - e_1)}{p_2' - p_1'} \qquad 40.22$$

$$T_v = \tfrac{1}{4}\pi U_z^2 \quad [U_z < 0.60] \qquad 40.23$$

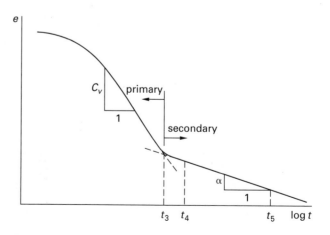

Figure 40.5 *Consolidation Parameters*

9. SECONDARY CONSOLIDATION

$$\alpha = \frac{-(e_5 - e_4)}{\log\left(\dfrac{t_5}{t_4}\right)} \qquad 40.24$$

$$C_\alpha = \frac{\alpha}{1 + e_o} \qquad 40.25$$

$$S_{\text{secondary}} = C_\alpha H \log_{10}\left(\frac{t_5}{t_4}\right) \qquad 40.26$$

10. SLOPE STABILITY IN SATURATED CLAY

$$d = \frac{D}{H} \qquad 40.27$$

$$F_{\text{cohesive}} = \frac{N_o c}{\gamma_{\text{eff}} H} \qquad 40.28$$

$$\gamma_{\text{eff}} = \gamma_{\text{saturated}} - \gamma_{\text{water}} \qquad 40.29$$

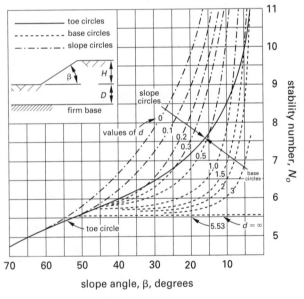

Source: *Soil Mechanics*, NAVFAC Design Manual DM-7.1, May 1982

Figure 40.6 *Taylor Slope Stability*
($\phi = 0°$)

11. LOADS ON BURIED PIPES

$$w = C\gamma B^2 \quad \left[\begin{array}{c}\text{Marston's formula;}\\ \text{rigid pipe}\end{array}\right] \qquad 40.30(b)$$

$$p = \frac{w}{B} \qquad 40.31$$

$$w = C\gamma BD \quad \text{[flexible pipe]} \qquad 40.32(b)$$

$$w = C_p \gamma D^2 \quad \text{[broad fill]} \qquad 40.33(b)$$

$$B_{\text{transition}} = D\sqrt{\frac{C_p}{C}} \qquad 40.34$$

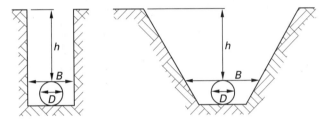

Figure 40.8 *Pipes in Backfilled Trenches*

12. ALLOWABLE PIPE LOADS

$$\text{crushing strength} = (\text{D-load strength})D_i \qquad 40.35$$

$$w_{\text{allowable}} = \left(\begin{array}{c}\text{known pipe}\\ \text{crushing strength}\end{array}\right)$$
$$\times \left(\frac{\text{LF}}{F}\right) \quad \text{[analysis]} \qquad 40.36$$

18. LIQUEFACTION

$$\frac{\tau_{h,\text{ave}}}{\sigma_o'} \approx 0.65\left(\frac{a_{\max}}{g}\right)\left(\frac{\sigma_o}{\sigma_o'}\right)r_d \qquad 40.37$$

Structural

CERM Chapter 41
Determinate Statics

Chapter, section, equation, figure, and table numbers correspond to CERM. For additional study material, go to the corresponding chapter and section number in CERM.

4. CONCENTRATED FORCES

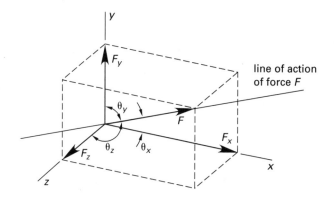

Figure 41.1 *Components and Direction Angles of a Force*

$$F_x = F \cos \theta_x \qquad 41.3$$

$$F_y = F \cos \theta_y \qquad 41.4$$

$$F_z = F \cos \theta_z \qquad 41.5$$

$$F = \sqrt{F_x^2 + F_y^2 + F_z^2} \qquad 41.6$$

6. MOMENT OF A FORCE ABOUT A POINT

$$\mathbf{M_O} = \mathbf{r} \times \mathbf{F} \qquad 41.7$$

$$M_O = |\mathbf{M_O}| = |\mathbf{r}|\,|\mathbf{F}| \sin \theta = d|\mathbf{F}| \quad [\theta \le 180°] \qquad 41.8$$

9. COMPONENTS OF A MOMENT

$$M_x = M \cos \theta_x \qquad 41.12$$

$$M_y = M \cos \theta_y \qquad 41.13$$

$$M_z = M \cos \theta_z \qquad 41.14$$

$$M_x = yF_z - zF_y \qquad 41.15$$

$$M_y = zF_x - xF_z \qquad 41.16$$

$$M_z = xF_y - yF_x \qquad 41.17$$

$$M = \sqrt{M_x^2 + M_y^2 + M_z^2} \qquad 41.18$$

10. COUPLES

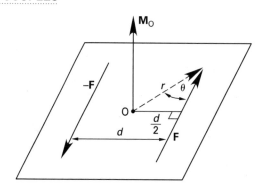

Figure 41.4 *Couple*

$$M_O = 2rF \sin \theta = Fd \qquad 41.19$$

15. MOMENT FROM A DISTRIBUTED LOAD

$$M_{\text{distributed load}} = \text{force} \times \text{distance}$$
$$= wx \left(\frac{x}{2} \right) = \tfrac{1}{2} wx^2 \qquad 41.34$$

17. CONDITIONS OF EQUILIBRIUM

$$\mathbf{F}_R = \sum \mathbf{F} = 0 \qquad 41.35$$

$$F_R = \sqrt{F_{R,x}^2 + F_{R,y}^2 + F_{R,z}^2} = 0 \qquad 41.36$$

$$\mathbf{M}_R = \sum \mathbf{M} = 0 \qquad 41.37$$

$$M_R = \sqrt{M_{R,x}^2 + M_{R,y}^2 + M_{R,z}^2} = 0 \qquad 41.38$$

25. INFLUENCE LINES FOR REACTIONS

$$R = F \times \text{influence line ordinate} \qquad 41.45$$

27. LEVERS

$$\begin{aligned}\frac{\text{mechanical}}{\text{advantage}} &= \frac{F_{\text{load}}}{F_{\text{applied}}} = \frac{\text{applied force lever arm}}{\text{load lever arm}} \\ &= \frac{\text{distance moved by applied force}}{\text{distance moved by load}} \qquad 41.46\end{aligned}$$

32. DETERMINATE TRUSSES

$$\text{no. of members} = 2(\text{no. of joints}) - 3 \quad 41.48$$

no. of members

+ no. of reactions

$$\begin{aligned}- 2(\text{no. of joints}) &= 0 \quad \text{[determinate]} \\ &> 0 \quad \text{[indeterminate]} \\ &< 0 \quad \text{[unstable]} \qquad 41.49\end{aligned}$$

40. PARABOLIC CABLES

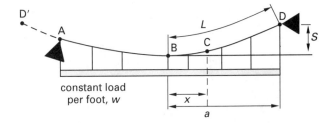

Figure 41.18 Parabolic Cable

$$\sum M_{\text{D}} = wa\left(\frac{a}{2}\right) - HS = 0 \qquad 41.50$$

$$H = \frac{wa^2}{2S} \qquad 41.51$$

$$T_{\text{C},x} = H = \frac{wa^2}{2S} \qquad 41.52$$

$$T_{\text{C},y} = wx \qquad 41.53$$

$$\begin{aligned}T_{\text{C}} &= \sqrt{\left(T_{\text{C},x}\right)^2 + \left(T_{\text{C},y}\right)^2} \\ &= w\sqrt{\left(\frac{a^2}{2S}\right)^2 + x^2} \qquad 41.54\end{aligned}$$

$$\tan\theta = \frac{wx}{H} \qquad 41.55$$

$$y(x) = \frac{wx^2}{2H} \qquad 41.56$$

$$L \approx a\left(1 + \left(\tfrac{2}{3}\right)\left(\frac{S}{a}\right)^2 - \left(\tfrac{2}{5}\right)\left(\frac{S}{a}\right)^4\right) \quad 41.57$$

42. CATENARY CABLES

$$y(x) = c\,\cosh\left(\frac{x}{c}\right) \qquad 41.64$$

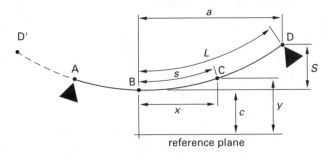

Figure 41.20 Catenary Cable

$$y = \sqrt{s^2 + c^2} = c\left(\cosh\left(\frac{x}{c}\right)\right) \qquad 41.65$$

$$s = c\left(\sinh\left(\frac{x}{c}\right)\right) \qquad 41.66$$

$$\text{sag} = S = y_{\text{D}} - c = c\left(\cosh\left(\frac{a}{c}\right) - 1\right) \quad 41.67$$

$$\tan\theta = \frac{s}{c} \qquad 41.68$$

$$H = wc \qquad 41.69$$

$$F = ws \qquad 41.70$$

$$T = wy \qquad 41.71$$

$$\tan\theta = \frac{ws}{H} \qquad 41.72$$

$$\cos\theta = \frac{H}{T} \qquad 41.73$$

CERM Chapter 42
Properties of Areas

> Chapter, section, equation, figure, and table numbers correspond to CERM. For additional study material, go to the corresponding chapter and section number in CERM.

1. CENTROID OF AN AREA

$$x_c = \frac{\int x\,dA}{A} \qquad 42.1$$

$$y_c = \frac{\int y\,dA}{A} \qquad 42.2$$

$$A = \int f(x)\,dx \qquad 42.3$$

$$x_c = \frac{\sum\limits_i A_i x_{ci}}{\sum\limits_i A_i}$$

42.5

$$y_c = \frac{\sum\limits_i A_i y_{ci}}{\sum\limits_i A_i}$$

42.6

2. FIRST MOMENT OF THE AREA

$$Q_y = \int x\, dA = x_c A$$

42.7

$$Q_x = \int y\, dA = y_c A$$

42.8

3. CENTROID OF A LINE

$$x_c = \frac{\int x\, dL}{L}$$

42.9

$$y_c = \frac{\int y\, dL}{L}$$

42.10

$$dL = \left(\sqrt{\left(\frac{dy}{dx}\right)^2 + 1} \right) dx$$

42.11

$$dL = \left(\sqrt{\left(\frac{dx}{dy}\right)^2 + 1} \right) dy$$

42.12

5. MOMENT OF INERTIA OF AN AREA

$$I_x = \int y^2\, dA$$

42.17

$$I_y = \int x^2\, dA$$

42.18

6. PARALLEL AXIS THEOREM

$$I_{\text{parallel axis}} = I_c + A d^2$$

42.20

7. POLAR MOMENT OF INERTIA

$$J = \int (x^2 + y^2)\, dA$$

42.21

$$J = I_x + I_y$$

42.22

$$J = I_{cx} + I_{cy}$$

42.23

8. RADIUS OF GYRATION

$$I = r^2 A$$

42.24

$$r = \sqrt{\frac{I}{A}}$$

42.25

$$r = \sqrt{\frac{J}{A}}$$

42.26

$$r^2 = r_x^2 + r_y^2$$

42.27

10. SECTION MODULUS

$$S = \frac{I_c}{c}$$

42.30

Centroids and Area Moments of Inertia for Basic Shapes

shape		centroidal location x_c	centroidal location y_c	area, A	area moment of inertia (rectangular and polar), I, J	radius of gyration, r
rectangle		$\dfrac{b}{2}$	$\dfrac{h}{2}$	bh	$I_x = \dfrac{bh^3}{3}$ $I_{c,x} = \dfrac{bh^3}{12}$ $J_c = \left(\dfrac{1}{12}\right)bh(b^2 + h^2)^*$	$r_x = \dfrac{h}{\sqrt{3}}$ $r_{c,x} = \dfrac{h}{2\sqrt{3}}$
triangular area			$\dfrac{h}{3}$	$\dfrac{bh}{2}$	$I_x = \dfrac{bh^3}{12}$ $I_{c,x} = \dfrac{bh^3}{36}$	$r_x = \dfrac{h}{\sqrt{6}}$ $r_{c,x} = \dfrac{h}{3\sqrt{2}}$
trapezoid			$h\left(\dfrac{b+2t}{3b+3t}\right)$	$\dfrac{(b+t)h}{2}$	$I_x = \dfrac{(b+3t)h^3}{12}$ $I_{c,x} = \dfrac{(b^2 + 4bt + t^2)h^3}{36(b+t)}$	$r_x = \left(\dfrac{h}{\sqrt{6}}\right)\sqrt{\dfrac{b+3t}{b+t}}$ $r_{c,x} = \dfrac{h\sqrt{2(b^2 + 4bt + t^2)}}{6(b+t)}$
circle		0	0	πr^2	$I_x = I_y = \dfrac{\pi r^4}{4}$ $J_c = \dfrac{\pi r^4}{2}$	$r_x = \dfrac{r}{2}$
quarter-circular area		$\dfrac{4r}{3\pi}$	$\dfrac{4r}{3\pi}$	$\dfrac{\pi r^2}{4}$	$I_x = I_y = \dfrac{\pi r^4}{16}$ $J_o = \dfrac{\pi r^4}{8}$	
semicircular area		0	$\dfrac{4r}{3\pi}$	$\dfrac{\pi r^2}{2}$	$I_x = I_y = \dfrac{\pi r^4}{8}$ $I_{c,x} = 0.1098r^4$ $J_o = \dfrac{\pi r^4}{4}$ $J_c = 0.5025r^4$	$r_x = \dfrac{r}{2}$ $r_{c,x} = 0.264r$
quarter-elliptical area		$\dfrac{4a}{3\pi}$	$\dfrac{4b}{3\pi}$	$\dfrac{\pi ab}{4}$	$I_x = \dfrac{\pi ab^3}{8}$ $I_y = \dfrac{\pi a^3 b}{8}$ $J_o = \dfrac{\pi ab(a^2 + b^2)}{8}$	
semielliptical area		0	$\dfrac{4b}{3\pi}$	$\dfrac{\pi ab}{2}$		
semiparabolic area		$\dfrac{3a}{8}$	$\dfrac{3h}{5}$	$\dfrac{2ah}{3}$		
parabolic area		0	$\dfrac{3h}{5}$	$\dfrac{4ah}{3}$	$I_x = \dfrac{4ah^3}{7}$ $I_y = \dfrac{4ha^3}{15}$ $I_{c,x} = \dfrac{16ah^3}{175}$	$r_x = h\sqrt{\dfrac{3}{7}}$ $r_y = \dfrac{a}{\sqrt{5}}$
parabolic spandrel	$y = kx^2$	$\dfrac{3a}{4}$	$\dfrac{3h}{10}$	$\dfrac{ah}{3}$	$I_x = \dfrac{ah^3}{21}$ $I_y = \dfrac{3ha^3}{15}$	
general spandrel	$y = kx^n$	$\left(\dfrac{n+1}{n+2}\right)a$	$\left(\dfrac{n+1}{4n+2}\right)h$	$\dfrac{ah}{n+1}$		
circular sector [α in radians]		$\dfrac{2r\sin\alpha}{3\alpha}$	0	αr^2		

*Theoretical definition based on $J = I_x + I_y$. However, in torsion, not all parts of the shape are effective. Effective values will be lower.

$$J = C\left(\dfrac{b^2 + h^2}{b^3 h^3}\right)$$

b/h	C
1	3.56
2	3.50
4	3.34
8	3.21

CERM Chapter 43
Material Properties and Testing

Chapter, section, equation, figure, and table numbers correspond to CERM. For additional study material, go to the corresponding chapter and section number in CERM.

1. TENSILE TEST

$$s = \frac{F}{A_o} \qquad 43.1$$

$$e = \frac{\delta}{L_o} \qquad 43.2$$

$$s = Ee \qquad 43.3$$

5. POISSON'S RATIO

$$\nu = \frac{e_{\text{lateral}}}{e_{\text{axial}}} = \frac{\dfrac{\Delta D}{D_o}}{\dfrac{\delta}{L_o}} \qquad 43.4$$

7. TRUE STRESS AND STRAIN

$$\sigma = \frac{F}{A} = \frac{F}{\left(1 - \dfrac{A_o - A}{A_o}\right) A_o} = \frac{F}{(1 - \text{RA})\, A_o}$$

$$= \frac{s}{(1 - \nu e)^2} = s(1 + e) \qquad 43.5$$

$$\epsilon = \int_{L_0}^{L} \frac{dL}{L} = \ln\left(\frac{L}{L_o}\right)$$
$$= \ln(1 + e) \quad [\text{prior to necking}] \qquad 43.6$$

$$A_o L_o = AL \qquad 43.7$$

$$\epsilon = \ln\left(\frac{A_o}{A}\right) = \ln\left(\frac{D_o}{D}\right)^2 = 2\ln\left(\frac{D_o}{D}\right) \qquad 43.8$$

$$\sigma = K\epsilon^n \qquad 43.9$$

8. DUCTILITY

$$\begin{aligned}\text{percent} \atop \text{elongation} &= \frac{L_f - L_o}{L_o} \times 100\% \\ &= e_f \times 100\% \qquad 43.10\end{aligned}$$

$$\text{ductility} = \frac{\text{ultimate failure strain}}{\text{yielding strain}} \qquad 43.11$$

$$\begin{aligned}\text{reduction} \atop \text{in area} &= \frac{A_o - A_f}{A_o} \times 100 \qquad 43.12\end{aligned}$$

9. STRAIN ENERGY

$$\text{work} = \text{force} \times \text{distance} = \int F\, dL \qquad 43.13$$

$$\begin{aligned}\text{work per} \atop \text{unit volume} &= \int \frac{F\, dL}{AL} = \int_0^{\epsilon_{\text{final}}} \sigma\, d\epsilon \qquad 43.14\end{aligned}$$

10. RESILIENCE

$$U_R = \int_0^{\epsilon_y} \sigma\, d\epsilon = E \int_0^{\epsilon_y} \epsilon\, d\epsilon = \frac{E\epsilon_y^2}{2} = \frac{S_y \epsilon_y}{2} \qquad 43.15$$

11. TOUGHNESS

$$U_T \approx S_u \epsilon_u \quad [\text{ductile}] \qquad 43.16$$

$$U_T \approx \left(\frac{S_y + S_u}{2}\right) \epsilon_u \quad [\text{ductile}] \qquad 43.17$$

$$U_T \approx \tfrac{2}{3} S_u \epsilon_u \quad [\text{brittle}] \qquad 43.18$$

14. TORSION TEST

$$\tau = G\theta \qquad 43.20$$

$$G = \frac{E}{2(1 + \nu)} \qquad 43.21$$

$$\gamma = \frac{TL}{JG} = \frac{\tau L}{rG} \quad [\text{radians}] \qquad 43.22$$

$$S_{ys} = \frac{S_{yt}}{\sqrt{3}} = 0.577 S_{yt} \qquad 43.23$$

CERM Chapter 44
Strength of Materials

Chapter, section, equation, figure, and table numbers correspond to CERM. For additional study material, go to the corresponding chapter and section number in CERM.

2. HOOKE'S LAW

$$\sigma = E\epsilon \qquad 44.2$$

$$\tau = G\phi \qquad 44.3$$

3. ELASTIC DEFORMATION

$$\delta = L_o \epsilon = \frac{L_o \sigma}{E} = \frac{L_o F}{EA} \qquad 44.4$$

$$L = L_o + \delta \qquad 44.5$$

4. TOTAL STRAIN ENERGY

$$U = \tfrac{1}{2}F\delta = \frac{F^2 L_o}{2AE} = \frac{\sigma^2 L_o A}{2E} \qquad 44.6$$

5. STIFFNESS AND RIGIDITY

$$k = \frac{F}{\delta} \quad \text{[general form]} \qquad 44.7(a)$$

$$k = \frac{AE}{L_o} \quad \text{[normal stress form]} \qquad 44.7(b)$$

$$R_j = \frac{k_j}{\displaystyle\sum_i k_i} \qquad 44.8$$

6. THERMAL DEFORMATION

$$\Delta L = \alpha L_o (T_2 - T_1) \qquad 44.9$$

$$\Delta A = \gamma A_o (T_2 - T_1) \qquad 44.10$$

$$\gamma \approx 2\alpha \qquad 44.11$$

$$\Delta V = \beta V_o (T_2 - T_1) \qquad 44.12$$

$$\beta \approx 3\alpha \qquad 44.13$$

$$\epsilon_{\text{th}} = \frac{\Delta L}{L_o} = \alpha(T_2 - T_1) \qquad 44.14$$

$$\sigma_{\text{th}} = E\epsilon_{\text{th}} \qquad 44.15$$

7. STRESS CONCENTRATIONS

$$\sigma' = K\sigma_0 \qquad 44.16$$

13. SHEAR STRESS IN BEAMS

$$\tau = \frac{V}{A} \qquad 44.28$$

$$\tau = \frac{V}{t_w d} \qquad 44.29$$

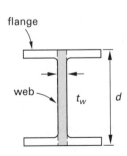

Figure 44.10 Web of a Flanged Beam

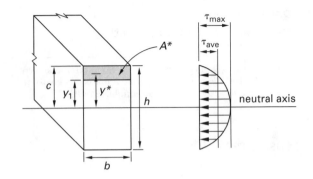

Figure 44.11 Shear Stress Distribution Within a Rectangular Beam

$$Q = y^* A^* \qquad 44.32$$

$$\tau_{\text{max,rectangular}} = \frac{3V}{2A} = \frac{3V}{2bh} \qquad 44.33$$

$$\tau_{\text{max,circular}} = \frac{4V}{3A} = \frac{4V}{3\pi r^2} \qquad 44.34$$

$$\tau_{\text{max,hollow cylinder}} = \frac{2V}{A} \qquad 44.35$$

14. BENDING STRESS IN BEAMS

$$\sigma_b = \frac{-My}{I_c} \qquad 44.36$$

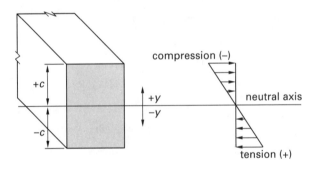

Figure 44.13 Bending Stress Distribution in a Beam

$$\sigma_{b,\text{max}} = \frac{Mc}{I_c} \qquad 44.37$$

$$\sigma_{b,\text{max}} = \frac{M}{S} \qquad 44.38$$

$$S = \frac{I_c}{c} \qquad 44.39$$

$$S_{\text{rectangular}} = \frac{bh^2}{6} \qquad 44.40$$

15. STRAIN ENERGY DUE TO BENDING MOMENT

$$U = \frac{1}{2EI} \int M^2(x)\, dx \qquad 44.41$$

16. ECCENTRIC LOADING OF AXIAL MEMBERS

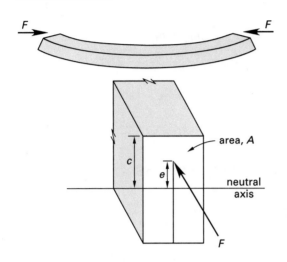

Figure 44.14 *Eccentric Loading of an Axial Member*

$$\sigma_{\text{max,min}} = \frac{F}{A} \pm \frac{Mc}{I_c} \qquad 44.42$$

$$= \frac{F}{A} \pm \frac{Fec}{I_c} \qquad 44.43$$

17. BEAM DEFLECTION: DOUBLE INTEGRATION METHOD

$$y = \text{deflection} \qquad 44.44$$

$$y' = \frac{dy}{dx} = \text{slope} \qquad 44.45$$

$$y'' = \frac{d^2y}{dx^2} = \frac{M(x)}{EI} \qquad 44.46$$

$$y''' = \frac{d^3y}{dx^3} = \frac{V(x)}{EI} \qquad 44.47$$

$$y = \frac{1}{EI} \int \left(\int M(x)\, dx \right) dx \qquad 44.48$$

18. BEAM DEFLECTION: MOMENT AREA METHOD

$$\phi = \int \frac{M(x)\, dx}{EI} \qquad 44.49$$

$$y = \int \frac{xM(x)\, dx}{EI} \qquad 44.50$$

19. BEAM DEFLECTION: STRAIN ENERGY METHOD

$$\tfrac{1}{2}Fy = \sum U \qquad 44.51$$

23. INFLECTION POINTS

$$y''(x) = \frac{1}{\rho(x)} = \frac{M(x)}{EI} \qquad 44.52$$

25. TRUSS DEFLECTION: VIRTUAL WORK METHOD

$$\delta = \sum \frac{SuL}{AE} \qquad 44.53$$

28. COMPOSITE STRUCTURES

$$n = \frac{E}{E_{\text{weakest}}} \qquad 44.54$$

$$\sigma_{\text{weakest}} = \frac{F}{A_t} \qquad 44.55$$

$$\sigma_{\text{stronger}} = \frac{nF}{A_t} \qquad 44.56$$

$$\sigma_{\text{weakest}} = \frac{Mc_{\text{weakest}}}{I_{c,t}} \qquad 44.57$$

$$\sigma_{\text{stronger}} = \frac{nMc_{\text{stronger}}}{I_{c,t}} \qquad 44.58$$

CERM Chapter 45
Basic Elements of Design

Chapter, section, equation, figure, and table numbers correspond to CERM. For additional study material, go to the corresponding chapter and section number in CERM.

1. SLENDER COLUMNS

$$F_e = \frac{\pi^2 EI}{L^2} = \frac{\pi^2 EA}{\left(\dfrac{L}{k}\right)^2} \qquad 45.1$$

$$\sigma_e = \frac{F_e}{A} = \frac{\pi^2 E}{\left(\dfrac{L}{k}\right)^2} \qquad 45.2$$

$$L' = CL \qquad 45.3$$

$$\sigma_e = \frac{F_e}{A} = \frac{\pi^2 E}{\left(\dfrac{CL}{k}\right)^2} \qquad 45.4$$

Table 45.1 *Theoretical End Restraint Coefficients*

illus.	end conditions	C ideal	recommended for design
(a)	both ends pinned	1	1.0*
(b)	both ends built in	0.5	0.65*–0.90
(c)	one end pinned, one end built in	0.707	0.80*–0.90
(d)	one end built in, one end free	2	2.0–2.1
(e)	one end built in, one end fixed against rotation but free	1	1.2*
(f)	one end pinned, one end fixed against rotation but free	2	2.0*

*AISC values

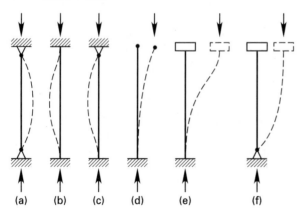

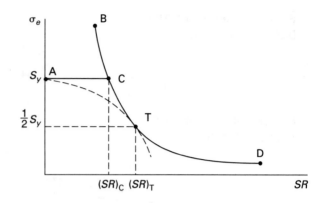

Figure 45.1 *Euler's Curve*

$$(\text{SR})_\text{T} = \sqrt{\frac{2\pi^2 CE}{S_y}} \qquad 45.5$$

2. INTERMEDIATE COLUMNS

$$\sigma_\text{cr} = \frac{P_\text{cr}}{A} = a - b\left(\frac{CL}{k}\right)^2 \qquad 45.6$$

$$\sigma_\text{cr} = S_y - \left(\frac{1}{E}\right)\left(\frac{S_y}{2\pi}\right)^2\left(\frac{CL}{k}\right)^2 \qquad 45.7$$

3. ECCENTRICALLY LOADED COLUMNS

$$\sigma_\text{max} = \sigma_\text{ave}(1 + \text{amplification factor})$$

$$= \left(\frac{F}{A}\right)\left(1 + \left(\frac{ec}{k^2}\right)\sec\left(\frac{\pi}{2}\sqrt{\frac{F}{F_e}}\right)\right)$$

$$= \left(\frac{F}{A}\right)\left(1 + \left(\frac{ec}{k^2}\right)\sec\left(\frac{L}{2k}\sqrt{\frac{F}{AE}}\right)\right)$$

$$= \left(\frac{F}{A}\right)\left(1 + \left(\frac{ec}{k^2}\right)\sec\phi\right) \qquad 45.8$$

$$\phi = \tfrac{1}{2}\left(\frac{L}{k}\right)\sqrt{\frac{F}{AE}} \qquad 45.9$$

4. THIN-WALLED CYLINDRICAL TANKS

$$\frac{t}{d_i} = \frac{t}{2r_i} < 0.1 \qquad \text{[thin-walled]} \qquad 45.10$$

$$\sigma_h = \frac{pr}{t} \qquad 45.11$$

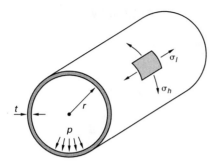

Figure 45.2 Stresses in a Thin-Walled Tank

$$\sigma_l = \frac{pr}{2t} \qquad \text{45.12}$$

$$\Delta L = L\epsilon_l$$

$$= L\left(\frac{\sigma_l - \nu\sigma_h}{E}\right) \qquad \text{45.13}$$

$$\Delta C = C\epsilon_h$$

$$= \pi d_o\left(\frac{\sigma_h - \nu\sigma_l}{E}\right) \qquad \text{45.14}$$

5. THICK-WALLED CYLINDERS

$$\sigma_c = \frac{r_i^2 p_i - r_o^2 p_o + \dfrac{(p_i - p_o)\, r_i^2 r_o^2}{r^2}}{r_o^2 - r_i^2} \qquad \text{45.15}$$

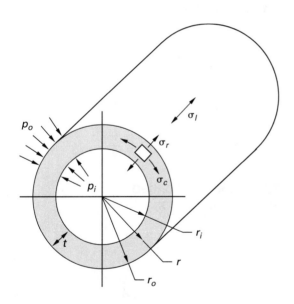

Figure 45.3 Thick-Walled Cylinder

$$\sigma_r = \frac{r_i^2 p_i - r_o^2 p_o - \dfrac{(p_i - p_o)\, r_i^2 r_o^2}{r^2}}{r_o^2 - r_i^2} \qquad \text{45.16}$$

$$\sigma_l = \frac{p_i r_i^2}{r_o^2 - r_i^2} \quad \begin{bmatrix} p_o \text{ does not act} \\ \text{longitudinally on the ends} \end{bmatrix} \quad \text{45.17}$$

Table 45.2 Stresses in Thick-Walled Cylinders[a]

stress	external pressure, p	internal pressure, p
$\sigma_{c,o}$	$\dfrac{-\left(r_o^2 + r_i^2\right) p}{r_o^2 - r_i^2}$	$\dfrac{2r_i^2 p}{r_o^2 - r_i^2}$
$\sigma_{r,o}$	$-p$	0
$\sigma_{c,i}$	$\dfrac{-2r_o^2 p}{r_o^2 - r_i^2}$	$\dfrac{\left(r_o^2 + r_i^2\right) p}{r_o^2 - r_i^2}$
$\sigma_{r,i}$	0	$-p$
τ_{max}	$\frac{1}{2}\sigma_{c,i}$	$\frac{1}{2}\left(\sigma_{c,i} + p\right)$

[a]Table 45.2 can be used with thin-walled cylinders. However, in most cases it will not be necessary to do so.

$$\epsilon = \frac{\Delta d}{d} = \frac{\Delta C}{C} = \frac{\Delta r}{r}$$

$$= \frac{\sigma_c - \nu(\sigma_r + \sigma_l)}{E} \qquad \text{45.18}$$

6. THIN-WALLED SPHERICAL TANKS

$$\sigma = \frac{pr}{2t} \qquad \text{45.19}$$

7. INTERFERENCE FITS

$$I_{diametral} = 2I_{radial}$$

$$= d_{o,inner} - d_{i,outer}$$

$$= |\Delta d_{o,inner}| + |\Delta d_{i,outer}| \qquad \text{45.20}$$

$$\epsilon = \frac{\Delta d}{d} = \frac{\Delta C}{C} = \frac{\Delta r}{r}$$

$$= \frac{\sigma_c - \nu\sigma_r}{E} \qquad \text{45.21}$$

$$I_{diametral} = 2I_{radial}$$

$$= \left(\frac{2pr_{o,shaft}}{E_{hub}}\right)\left(\frac{r_{o,hub}^2 + r_{o,shaft}^2}{r_{o,hub}^2 - r_{o,shaft}^2} + \nu_{hub}\right)$$

$$+ \left(\frac{2pr_{o,shaft}}{E_{shaft}}\right)\left(\frac{r_{o,shaft}^2 + r_{i,shaft}^2}{r_{o,shaft}^2 - r_{i,shaft}^2} - \nu_{shaft}\right) \qquad \text{45.22}$$

$$I_{diametral} = 2I_{radial}$$

$$= \left(\frac{4pr_{shaft}}{E}\right)\left(\frac{1}{1 - \left(\dfrac{r_{shaft}}{r_{o,hub}}\right)^2}\right) \qquad \text{45.23}$$

$$F_{\max} = fN = 2\pi f p r_{o,\text{shaft}} L_{\text{interface}} \qquad 45.24$$

$$T_{\max} = 2\pi f p r_{o,\text{shaft}}^2 L_{\text{interface}} \qquad 45.25$$

10. RIVET AND BOLT CONNECTIONS

$$\tau = \frac{F}{\frac{\pi}{4}d^2} \qquad 45.26$$

$$A_t = t(b - nd) \qquad 45.28$$

$$\sigma_t = \frac{F}{A_t} \qquad 45.29$$

$$\sigma_p = \frac{F}{dt} \qquad 45.30$$

$$n = \frac{\sigma_p}{\text{allowable bearing stress}} \qquad 45.31$$

$$\tau = \frac{F}{2A} = \frac{F}{2t\left(L - \dfrac{d}{2}\right)} \qquad 45.32$$

11. BOLT PRELOAD

$$k_{\text{bolt}} = \frac{F}{\Delta L} = \frac{A_{\text{bolt}} E_{\text{bolt}}}{L_{\text{bolt}}} \qquad 45.33$$

$$k_{\text{parts}} = \frac{A_{e,\text{parts}} E_{\text{parts}}}{L_{\text{parts}}} \qquad 45.34$$

$$\frac{1}{k_{\text{parts,composite}}} = \frac{1}{k_1} + \frac{1}{k_2} + \frac{1}{k_3} + \cdots \qquad 45.35$$

$$F_{\text{bolt}} = F_i + \frac{k_{\text{bolt}} F_{\text{applied}}}{k_{\text{bolt}} + k_{\text{parts}}} \qquad 45.36$$

$$F_{\text{parts}} = \frac{k_{\text{parts}} F_{\text{applied}}}{k_{\text{bolt}} + k_{\text{parts}}} - F_i \qquad 45.37$$

$$\sigma_{\text{bolt}} = \frac{KF}{A} \qquad 45.38$$

12. BOLT TORQUE TO OBTAIN PRELOAD

$$T = K_T d_{\text{bolt}} F_i \qquad 45.39$$

$$K_T = \frac{f_c r_c}{d_{\text{bolt}}} + \left(\frac{r_t}{d_{\text{bolt}}}\right)\left(\frac{\tan\theta + f_t \sec\alpha}{1 - f_t \tan\theta \sec\alpha}\right) \qquad 45.40$$

$$\tan\theta = \frac{\text{lead per revolution}}{2\pi r_t} \qquad 45.41$$

13. FILLET WELDS

$$t_e = 0.707y \qquad 45.42$$

$$\tau = \frac{F}{bt_e} \qquad 45.43$$

14. CIRCULAR SHAFT DESIGN

$$\tau = G\theta = \frac{Tr}{J} \qquad 45.44$$

$$U = \frac{T^2 L}{2GJ} \qquad 45.45$$

$$J = \frac{\pi r^4}{2} = \frac{\pi d^4}{32} \qquad 45.46$$

$$J = \frac{\pi}{2}\left(r_o^4 - r_i^4\right) \qquad 45.47$$

$$\gamma = \frac{L\theta}{r} = \frac{TL}{GJ} \qquad 45.48$$

$$G = \frac{E}{2(1 + \nu)} \qquad 45.49$$

$$T_{\text{in-lbf}} = \frac{(63{,}025)(\text{horsepower})}{n_{\text{rpm}}} \qquad 45.50$$

$$\tau_{\max} = \sqrt{\left(\frac{\sigma_x}{2}\right)^2 + \tau^2} \qquad 45.51$$

$$\tau_{\max} = \frac{16}{\pi d^3}\sqrt{M^2 + T^2} \qquad 45.52$$

$$\sigma' = \frac{16}{\pi d^3}\sqrt{4M^2 + 3T^2} \qquad 45.53$$

15. TORSION IN THIN-WALLED, NONCIRCULAR SHELLS

$$\tau = \frac{T}{2At} \qquad 45.54$$

$$q = \tau t = \frac{T}{2A} \quad [\text{constant}] \qquad 45.55$$

$$\phi = \frac{TLp}{4A^2 tG} \qquad 45.56$$

18. ECCENTRICALLY LOADED BOLTED CONNECTIONS

$$\tau = \frac{Tr}{J} = \frac{Fer}{J} \qquad 45.57$$

$$J = \sum_i r_i^2 A_i \qquad 45.58$$

$$\tau_v = \frac{F}{nA} \qquad 45.59$$

21. SPRINGS

$$F = k\delta \qquad 45.60$$

$$k = \frac{F_1 - F_2}{\delta_1 - \delta_2} \qquad 45.61$$

$$\Delta E_p = \tfrac{1}{2}k\delta^2 \qquad 45.62$$

$$m\left(\frac{g}{g_c}\right)(h + \delta) = \tfrac{1}{2}k\delta^2 \qquad 45.63(b)$$

$$\frac{1}{k_{eq}} = \frac{1}{k_1} + \frac{1}{k_2} + \frac{1}{k_3} + \cdots \qquad 45.64$$

$$k_{eq} = k_1 + k_2 + k_3 + \cdots \qquad 45.65$$

22. WIRE ROPE

$$\sigma_{bending} = \frac{d_w E_w}{d_{sh}} \qquad 45.66$$

$$p_{bearing} = \frac{2F_t}{d_r d_{sh}} \qquad 45.67$$

CERM Chapter 46
Structural Analysis I

Chapter, section, equation, figure, and table numbers correspond to CERM. For additional study material, go to the corresponding chapter and section number in CERM.

4. REVIEW OF ELASTIC DEFORMATION

$$\delta = \frac{FL}{AE} \qquad 46.2$$

$$\delta = \alpha L_o \Delta T \qquad 46.3$$

7. THREE-MOMENT EQUATION

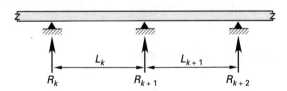

Figure 46.2 Portion of a Continuous Beam

$$\frac{M_k L_k}{I_k} + (2M_{k+1})\left(\frac{L_k}{I_k} + \frac{L_{k+1}}{I_{k+1}}\right) + \frac{M_{k+2}L_{k+1}}{I_{k+1}}$$
$$= -6\left(\frac{A_k a}{I_k L_k} + \frac{A_{k+1}b}{I_{k+1}L_{k+1}}\right) \qquad 46.4$$

8. FIXED END MOMENTS

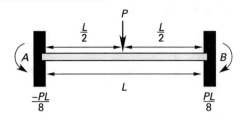

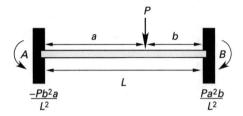

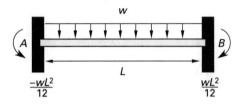

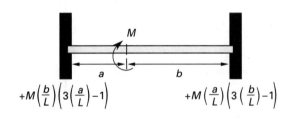

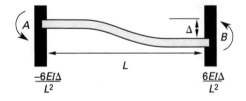

10. INFLUENCE DIAGRAMS

$$R_L = P \times \text{ordinate} \qquad 46.9$$

$$R_L = \int_{x_1}^{x_2}(w \times \text{ordinate})dx$$
$$= w \times \text{area under curve} \qquad 46.10$$

CERM Chapter 47
Structural Analysis II

> Chapter, section, equation, figure, and table numbers correspond to CERM. For additional study material, go to the corresponding chapter and section number in CERM.

3. REVIEW OF WORK AND ENERGY

$$W = P\Delta \quad \text{[linear displacement]} \qquad 47.1$$

$$W = T\phi \quad \text{[rotation]} \qquad 47.2$$

$$W = U_2 - U_1 \qquad 47.3$$

4. REVIEW OF LINEAR DEFORMATION

$$\Delta = \frac{PL}{AE} \qquad 47.4$$

5. THERMAL LOADING

$$P_{\text{th}} = \frac{\Delta_{\text{constrained}} AE}{L} = \alpha(T_2 - T_1)\left(\frac{LAE}{L}\right)$$
$$= \alpha(T_2 - T_1)AE \qquad 47.5$$

$$M = \alpha(T_{\text{extreme fiber}} - T_{\text{neutral axis}})\left(\frac{EI}{\frac{h}{2}}\right) \qquad 47.6$$

6. DUMMY UNIT LOAD METHOD

$$W_Q = 1 \times \Delta = \int \sigma_Q \epsilon_P dV \qquad 47.7$$

$$W_Q = W_m + W_v + W_a + W_t \qquad 47.8$$

$$W_f = \int \left(\frac{m_Q m_P}{EI}\right) ds \qquad 47.9$$

$$W_v = \int \left(\frac{V_Q V_P}{GA}\right) ds \qquad 47.10$$

$$W_a = \int \left(\frac{N_Q N_P}{EA}\right) ds \qquad 47.11$$

$$W_t = \int \left(\frac{T_Q T_P}{GJ}\right) ds \qquad 47.12$$

7. BEAM DEFLECTIONS BY THE DUMMY UNIT LOAD METHOD

$$W_Q = W_f = \int \left(\frac{m_Q m_P}{EI}\right) ds \qquad 47.13$$

23. APPROXIMATE METHOD: MOMENT COEFFICIENTS

$$M = C_1 w L^2 \qquad 47.42$$

24. APPROXIMATE METHOD: SHEAR COEFFICIENTS

$$V = C_2 \left(\frac{wL}{2}\right) \qquad 47.43$$

condition	C_1
positive moments near mid-span	
end spans	
simple support	$\frac{1}{11}$
built in	$\frac{1}{14}$
interior spans	$\frac{1}{16}$
negative moments at exterior face of first interior support	
two spans	$\frac{1}{9}$
three or more spans	$\frac{1}{10}$
negative moments at other faces of interior supports	$\frac{1}{11}$
negative moments at exterior built-in supports	
support is a column	$\frac{1}{16}$
support is a cross beam or girder	$\frac{1}{24}$

C_2 for shear in end members at the first interior support is 1.15. For shear at the face of all other supports, $C_2 = 1.0$.

CERM Chapter 48
Properties of Concrete and Reinforcing Steel

> Chapter, section, equation, figure, and table numbers correspond to CERM. For additional study material, go to the corresponding chapter and section number in CERM.

8. COMPRESSIVE STRENGTH

$$f_c' = \frac{P}{A}$$

10. MODULUS OF ELASTICITY

$$E_c = w_c^{1.5} 33\sqrt{f_c'} \quad \left[90 \frac{\text{lbf}}{\text{ft}^3} \le w_c \le 155 \frac{\text{lbf}}{\text{ft}^3}\right] \qquad 48.1(b)$$

$$E_c = 57{,}000\sqrt{f_c'} \quad \text{[normal-weight concrete]} \qquad 48.2(b)$$

11. SPLITTING TENSILE STRENGTH

$$f_{ct} = \frac{2P}{\pi DL} \qquad 48.3$$

$$f_{ct} = 6.7\sqrt{f_c'} \quad \text{[normal weight]} \qquad 48.4$$

$$f_{ct} = 5.7\sqrt{f_c'} \quad \text{[sand lightweight]} \qquad 48.5$$

$$f_{ct} = 5\sqrt{f_c'} \quad \text{[all lightweight]} \qquad 48.6$$

12. MODULUS OF RUPTURE

$$f_r = \frac{Mc}{I} \quad \text{[tension]} \qquad 48.7$$

$$f_r = 7.5\sqrt{f_c'} \qquad 48.8(b)$$

CERM Chapter 49
Concrete Proportioning, Mixing, and Placing

> Chapter, section, equation, figure, and table numbers correspond to CERM. For additional study material, go to the corresponding chapter and section number in CERM.

6. ABSOLUTE VOLUME METHOD

$$V_{\text{absolute}} = \frac{W}{(\text{SG})\gamma_{\text{water}}} \qquad 49.1(b)$$

Table 49.2 *Summary of Approximate Properties of Concrete Components*

cement	
specific weight	195 lbf/ft^3
specific gravity	3.13–3.15
weight of one sack	94 lbf
fine aggregate	
specific weight	165 lbf/ft^3
specific gravity	2.64
coarse aggregate	
specific weight	165 lbf/ft^3
specific gravity	2.64
water	
specific weight	62.4 lbf/ft^3
	7.48 gal/ft^3
	8.34 lbf/gal
	239.7 gal/ton
specific gravity	1.00

(Multiply lbf/ft^3 by 16 to obtain kg/m^3.)
(Multiply lbf by 0.45 to obtain kg.)

Table 49.3 *Dry and Wet Basis Calculations*

	dry basis	wet basis
fraction moisture, f	$\dfrac{W_{\text{excess water}}}{W_{\text{SSD sand}}}$	$\dfrac{W_{\text{excess water}}}{W_{\text{SSD sand}} + W_{\text{excess water}}}$
weight of sand, $W_{\text{wet sand}}$	$W_{\text{SSD sand}} + W_{\text{excess water}}$ $(1+f)W_{\text{SSD sand}}$	$W_{\text{SSD sand}} + W_{\text{excess water}}$ $\left(\dfrac{1}{1-f}\right)W_{\text{SSD sand}}$
weight of SSD sand, $W_{\text{SSD sand}}$	$\dfrac{W_{\text{wet sand}}}{1+f}$	$(1-f)W_{\text{wet sand}}$
weight of excess water, $W_{\text{excess water}}$	$fW_{\text{SSD sand}}$ $\dfrac{fW_{\text{wet sand}}}{1+f}$	$fW_{\text{wet sand}}$ $\dfrac{fW_{\text{SSD sand}}}{1-f}$

CERM Chapter 50
Reinforced Concrete: Beams

Chapter, section, equation, figure, and table numbers correspond to CERM. For additional study material, go to the corresponding chapter and section number in CERM.

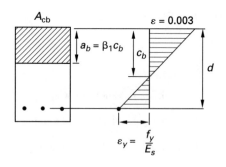

Figure 50.3 Beam at Balanced Condition

5. SERVICE LOADS, FACTORED LOADS, AND LOAD COMBINATIONS

$$U = 1.4D + 1.7L \qquad 50.1$$

$$U = 0.75(1.4D + 1.7L + 1.7W) \qquad 50.2$$

$$U = 0.9D + 1.3W \qquad 50.3$$

9. STEEL COVER AND BEAM WIDTH

$$1.5 \leq d/b \leq 2.5 \qquad 50.14$$

6. DESIGN STRENGTH AND DESIGN CRITERIA

$$\text{design strength} = (\text{nominal strength})\phi \qquad 50.4$$

$$\phi = 0.9 \quad [\text{flexure}] \qquad 50.5(a)$$

$$\phi = 0.85 \quad [\text{for shear}] \qquad 50.5(b)$$

$$\phi M_n \geq M_u \qquad 50.6(a)$$

$$\phi V_n \geq V_u \qquad 50.6(b)$$

Table 50.2 Minimum Beam Widths[a,b]
(inches)

size of bar	number of bars in a single layer of reinforcement							add for each additional bar[c]
	2	3	4	5	6	7	8	
no. 4	6.1	7.6	9.1	10.6	12.1	13.6	15.1	1.50
no. 5	6.3	7.9	9.6	11.2	12.8	14.4	16.1	1.63
no. 6	6.5	8.3	10.0	11.8	13.5	15.3	17.0	1.75
no. 7	6.7	8.6	10.5	12.4	14.2	16.1	18.0	1.88
no. 8	6.9	8.9	10.9	12.9	14.9	16.9	18.9	2.00
no. 9	7.3	9.5	11.8	14.0	16.3	18.6	20.8	2.26
no. 10	7.7	10.2	12.8	15.3	17.8	20.4	22.9	2.54
no. 11	8.0	10.8	13.7	16.5	19.3	22.1	24.9	2.82
no. 14	8.9	12.3	15.6	19.0	22.4	25.8	29.2	3.39
no. 18	10.5	15.0	19.5	24.0	28.6	33.1	37.6	4.51

(Multiply in by 25.4 to obtain mm.)
[a]Using no. 3 stirrups. If stirrups are not used, deduct 0.75 in.
[b]The bar diameter for no. 4 bars and larger is less than the minimum bend diameter for $1\frac{1}{2}$ in for no. 3 stirrups. A small allowance has been included in the beam widths to achieve a full bend radius.
[c]For additional horizontal bars, the beam width is increased by adding the value in the last column.

7. MINIMUM STEEL AREA

$$A_{s,\min} = \frac{3\sqrt{f_c'}\,b_w d}{f_y} \geq \frac{200 b_w d}{f_y} \qquad 50.7(b)$$

$$A_{s,\min} = \frac{6\sqrt{f_c'}\,b_w d}{f_y} \qquad 50.8(b)$$

8. MAXIMUM STEEL AREA

$$A_{s,\max} = 0.75 A_{\mathrm{sb}} \qquad 50.9$$

$$A_{\mathrm{sb}} = \frac{0.85 f_c' A_{\mathrm{cb}}}{f_y} \qquad 50.10$$

$$a_b = \beta_1 \left(\frac{87{,}000}{87{,}000 + f_y} \right) d \qquad 50.11(b)$$

$$\beta_1 = 0.85 \quad [f_c' \leq 4000 \text{ lbf/in}^2] \qquad 50.12(b)$$

$$\beta_1 = 0.85 - (0.05)\left(\frac{f_c' - 4000}{1000} \right) \geq 0.65 \qquad 50.12(d)$$

$$A_{\mathrm{sb}} = \left(\left(\frac{0.85 \beta_1 f_c'}{f_y} \right) \left(\frac{87{,}000}{87{,}000 + f_y} \right) \right) bd \qquad 50.13(b)$$

10. NOMINAL MOMENT CAPACITY OF SINGLY REINFORCED SECTIONS

$$M_n = f_y A_s \times \text{lever arm} \qquad 50.15$$

$$T = f_y A_s \qquad 50.16$$

$$A_c = \frac{T}{0.85 f_c'} = \frac{f_y A_s}{0.85 f_c'} \qquad 50.17$$

$$M_n = A_s f_y (d - \lambda) \qquad 50.18$$

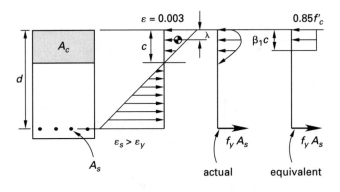

Figure 50.4 *Conditions at Maximum Moment*

11. BEAM DESIGN: SIZE KNOWN, REINFORCEMENT UNKNOWN

$$A_s = \frac{M_u}{\phi f_y (d - \lambda)} \qquad \text{50.19}$$

12. BEAM DESIGN: SIZE AND REINFORCEMENT UNKNOWN

$$\lambda = \frac{A_c}{2b} \qquad \text{50.20}$$

$$M_n = A_s f_y d \left(1 - \frac{A_s f_y}{1.7 f_c' b d} \right) \qquad \text{50.21}$$

$$\rho = \frac{A_s}{bd} \qquad \text{50.22}$$

$$M_n = \rho b d^2 f_y \left(1 - \frac{\rho f_y}{1.7 f_c'} \right)$$

$$= A_s f_y \left(d - \frac{A_s f_y}{1.7 f_c' b} \right) \qquad \text{50.23}$$

$$\rho_{sb} = \left(\frac{0.85 \beta_1 f_c'}{f_y} \right) \left(\frac{87,000}{87,000 + f_y} \right)$$
$$\qquad \text{50.24(b)}$$

$$\rho_{\min} = \frac{3 \sqrt{f_c'}}{f_y} \geq \frac{200}{f_y} \qquad \text{50.25(b)}$$

Table 50.3 *Total Areas for Various Numbers of Bars*

bar size	nominal diameter (in)	weight (lbf/ft)	number of bars									
			1	2	3	4	5	6	7	8	9	10
no. 3	0.375	0.376	0.11	0.22	0.33	0.44	0.55	0.66	0.77	0.88	0.99	1.10
no. 4	0.500	0.668	0.20	0.40	0.60	0.80	1.00	1.20	1.40	1.60	1.80	2.00
no. 5	0.625	1.043	0.31	0.62	0.93	1.24	1.55	1.86	2.17	2.48	2.79	3.10
no. 6	0.750	1.502	0.44	0.88	1.32	1.76	2.20	2.64	3.08	3.52	3.96	4.40
no. 7	0.875	2.044	0.60	1.20	1.80	2.40	3.00	3.60	4.20	4.80	5.40	6.00
no. 8	1.000	2.670	0.79	1.58	2.37	3.16	3.95	4.74	5.53	6.32	7.11	7.90
no. 9	1.128	3.400	1.00	2.00	3.00	4.00	5.00	6.00	7.00	8.00	9.00	10.0
no. 10	1.270	4.303	1.27	2.54	3.81	5.08	6.35	7.62	8.89	10.16	11.43	12.70
no. 11	1.410	5.313	1.56	3.12	4.68	6.24	7.80	9.36	10.92	12.48	14.04	15.60
no. 14[a]	1.693	7.650	2.25	4.50	6.75	9.00	11.25	13.5	15.75	18.00	20.25	22.50
no. 18[a]	2.257	13.60	4.00	8.00	12.0	16.0	20.00	24.0	28.00	32.00	36.00	40.00

(Multiply in^2 by 645 to obtain mm^2.)
[a] No. 14 and no. 18 are typically used in columns only.

13. SERVICEABILITY: CRACKING

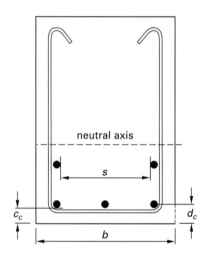

Figure 50.5 *Parameters for Cracking Calculation*

$$s \le \frac{540}{f_s} - 2.5 c_c \le \frac{432}{f_s} \qquad 50.27$$

14. CRACKED MOMENT OF INERTIA

$$n = \frac{E_s}{E_c} \qquad 50.28$$

$$E_c = 57{,}000\sqrt{f_c'} \qquad 50.29(b)$$

$$\frac{b c_s^2}{2} = n A_s (d - c_s) \qquad 50.30$$

$$c_s = \left(\frac{n A_s}{b}\right)\left(\sqrt{1 + \frac{2bd}{n A_s}} - 1\right) \qquad 50.31$$

$$c_s = n\rho d\left(\sqrt{1 + \frac{2}{n\rho}} - 1\right) \qquad 50.32$$

$$I_{cr} = \frac{b c_s^3}{3} + n A_s (d - c_s)^2 \qquad 50.33$$

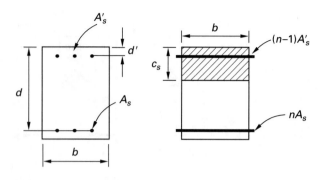

(a) reinforced section   (b) transformed section

Figure 50.6 *Parameters for Cracked Moment of Inertia*

15. SERVICEABILITY: DEFLECTIONS

$$I_e = \left(\frac{M_{cr}}{M_a}\right)^3 I_g + \left(1 - \left(\frac{M_{cr}}{M_a}\right)^3\right) I_{cr} \le I_g \qquad 50.34$$

$$M_{cr} = \frac{f_r I_g}{y_t} \qquad 50.35$$

$$f_r = 7.5\sqrt{f_c'} \qquad 50.36(b)$$

16. LONG-TERM DEFLECTIONS

$$\Delta_a = \lambda \Delta_i \qquad 50.37$$

$$\lambda = \frac{i}{1 + 50\rho'} \qquad 50.38$$

17. MINIMUM BEAM DEPTHS TO AVOID EXPLICIT DEFLECTION CALCULATIONS

Table 50.4 *Minimum Beam Thickness Unless Deflections are Computed ($f_y = 60{,}000 \text{ lbf/in}^2$)* [a]

construction	minimum h (fraction of span length)
simply supported	$\frac{1}{16}$
one end continuous	$\frac{1}{18.5}$
both ends continuous	$\frac{1}{21}$
cantilever	$\frac{1}{8}$

[a] Corrections are required for other steel strengths. See Eq. 50.43.
Source: ACI 318 Table 9.5(a)

$$0.4 + \frac{f_y}{100{,}000} \qquad 50.39(b)$$

$$1.65 - 0.005w \ge 1.09 \qquad 50.40(b)$$

19. DESIGN OF T-BEAMS

Case 1: Beams with flanges on each side of the web [ACI 318 8.10.2]

Effective width (including the compression area of the stem) for a T-beam is the minimum of

> one fourth of the beam's span length, or
>
> the stem width plus 16 times the thickness of the slab, or
>
> the beam spacing

Case 2: Beams with an L-shaped flange [ACI 318 8.10.3]

Effective width (including the compression area of the stem) is the minimum of

> the stem width plus one twelfth of the beam's span length, or
>
> the stem width plus 6 times the thickness of the slab, or
>
> the stem width plus one half of the clear distance between beam webs

21. SHEAR REINFORCEMENT

$$\phi V_n = V_u \qquad 50.41$$

$$V_n = V_c + V_s \qquad 50.42$$

22. SHEAR STRENGTH PROVIDED BY CONCRETE

$$V_c = \left(1.9\sqrt{f_c'} + 2500\rho_w \left(\frac{V_u d}{M_u}\right)\right) b_w d \leq 3.5\sqrt{f_c'} b_w d$$
$$\qquad 50.43(b)$$

$$V_c = 2\sqrt{f_c'} b_w d \qquad 50.44(b)$$

23. SHEAR STRENGTH PROVIDED BY SHEAR REINFORCEMENT

$$V_s = \frac{A_v f_y d}{s} \qquad 50.45$$

$$V_s = \frac{A_v f_y (\sin\theta + \cos\theta) d}{s} \qquad 50.46$$

$$V_s = A_v f_y \sin\theta \leq 3\sqrt{f_c'} b_w d \qquad 50.47(b)$$

24. SHEAR REINFORCEMENT LIMITATIONS

$$V_s \leq 8\sqrt{f_c'} b_w d \qquad 50.48(b)$$

$$V_{u,\max} = 10\phi\sqrt{f_c'} b_w d \qquad 50.49(b)$$

25. STIRRUP SPACING

$$s = \min\{24 \text{ in or } d/2\} \quad [V_s \leq 4\sqrt{f_c'} b_w d]$$
$$\qquad 50.50(b)$$

$$s = \min\{12 \text{ in or } d/2\} \quad [V_s > 4\sqrt{f_c'} b_w d]$$
$$\qquad 50.51(b)$$

$$A_{v,\min} = \frac{50 b_w s}{f_y} \qquad 50.52(b)$$

26. NO-STIRRUP CONDITIONS

$$\frac{\phi V_c}{2} \geq V_u \qquad 50.53$$

27. SHEAR REINFORCEMENT DESIGN PROCEDURE

$$V_{s,\text{req}} = \frac{V_u}{\phi} - V_c \qquad 50.54$$

$$s = \frac{A_v f_y d}{V_{s,\text{req}}} \qquad 50.55$$

28. ANCHORAGE OF SHEAR REINFORCEMENT

$$\text{embedment length} = \frac{0.014 d_b f_y}{\sqrt{f_c'}} \qquad 50.56(b)$$

30. STRENGTH ANALYSIS OF DOUBLY REINFORCED SECTIONS

$$A_c = \frac{f_y A_s - f_s' A_s'}{0.85 f_c'} \qquad 50.57$$

$$\varepsilon_s' = \left(\frac{0.003}{c}\right)(c - d') \qquad 50.58$$

$$f_s' = E_s \varepsilon_s' \leq f_y \qquad 50.59$$

$$M_n = A_s f_y (d - \lambda) + A_s' f_s' (d - d') \qquad 50.60$$

31. DESIGN OF DOUBLY REINFORCED SECTIONS

$$M_n' = \frac{M_u}{\phi} - M_{nc} \qquad 50.61$$

$$A_{s,\text{add}} = \frac{M_n'}{f_y (d - d')} \qquad 50.62$$

$$A_s = A_{s,\max} + A_{s,\text{add}} \qquad 50.63$$

$$c = 0.75 \left(\frac{87,000}{87,000 + f_y}\right) d \qquad 50.64(b)$$

$$A_s' = \left(\frac{f_y}{f_s'}\right) A_{s,\text{add}} \qquad 50.65$$

32. DEEP BEAMS

$$l_c = \min\{0.15l_n, d\} \qquad \textit{50.66}$$

$$l_c = \min\{0.5a, d\} \qquad \textit{50.67}$$

$$V_s = \left(\left(\frac{A_v}{s} \right) \left(\frac{1 + \frac{l_n}{d}}{12} \right) \right.$$
$$\left. + \left(\frac{A_{vh}}{s_2} \right) \left(\frac{11 - \frac{l_n}{d}}{12} \right) \right) f_y d \qquad \textit{50.68}$$

$$A_v \geq 0.0015 b_w s \qquad \textit{50.69}$$

$$s \leq \min\{d/5, 18 \text{ in}\} \qquad \textit{50.70(b)}$$

$$A_{vh} \geq 0.0025 b_w s_2 \qquad \textit{50.71}$$

$$s_2 \leq \min\{d/3, 18 \text{ in}\} \qquad \textit{50.72(b)}$$

$$V_n \leq 8\sqrt{f_c'} b_w d \quad [l_n/d \leq 2] \qquad \textit{50.73(b)}$$

$$V_n = \tfrac{2}{3}\left(10 + \frac{l_n}{d}\right)\sqrt{f_c'} b_w d \quad [2 \leq l_n/d \leq 5]$$
$$\textit{50.74(b)}$$

CERM Chapter 51
Reinforced Concrete: Slabs

> Chapter, section, equation, figure, and table numbers correspond to CERM. For additional study material, go to the corresponding chapter and section number in CERM.

3. TEMPERATURE STEEL

$$\rho_t = (0.0018)\left(\frac{60{,}000}{f_y} \right) \qquad \textit{51.1(b)}$$

6. SLAB DESIGN FOR FLEXURE

$$s = \frac{A_b}{A_{sr}} \qquad \textit{51.2}$$

9. DIRECT DESIGN METHOD

$$M_o = \frac{w_u l_2 l_n^2}{8} \qquad \textit{51.3}$$

10. FACTORED MOMENTS IN SLAB BEAMS

$$\alpha = \frac{E_{cb} I_b}{E_{cs} I_s} \qquad \textit{51.4}$$

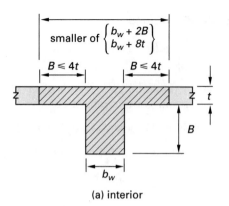

Figure 51.4 *Two-Way Slab Beams (monolithic or fully composite construction) [ACI 318 13.2.4]*

11. COMPUTATION OF β_t

$$\beta_t = \frac{E_{cb} C}{2 E_{cs} I_s} \qquad \textit{51.5}$$

$$C = \sum \left(1 - 0.63\left(\frac{x}{y} \right) \right)\left(\frac{x^3 y}{3} \right) \qquad \textit{51.6}$$

12. DEFLECTIONS IN TWO-WAY SLABS

$$h = \frac{l_n\left(0.8 + \dfrac{f_y}{200{,}000} \right)}{36 + 5\beta(\alpha_m - 0.2)}$$
$$\geq 5 \text{ in} \qquad \textit{51.7}$$

$$h = \frac{l_n\left(0.8 + \dfrac{f_y}{200{,}000} \right)}{36 + 9\beta}$$
$$\geq 3.5 \text{ in} \qquad \textit{51.8}$$

CERM Chapter 52
Reinforced Concrete: Short Columns

Chapter, section, equation, figure, and table numbers correspond to CERM. For additional study material, go to the corresponding chapter and section number in CERM.

1. INTRODUCTION

$$\frac{k_b l_u}{r} \leq 34 - 12 \left(\frac{M_1}{M_2}\right) \qquad 52.1$$

$$\frac{k_u l_u}{r} \leq 22 \quad [k_u > 1] \qquad 52.2$$

2. TIED COLUMNS

$$0.01 \leq \rho_g \leq 0.08 \qquad 52.3$$

3. SPIRAL COLUMNS

$$\rho_s = 0.45 \left(\frac{A_g}{A_c} - 1\right) \left(\frac{f'_c}{f_y}\right) \qquad 52.4$$

$$s \approx \frac{4 A_{sp}}{\rho_s D_c} \qquad 52.5$$

$$\text{clear distance} = s - d_{sp} \qquad 52.6$$

4. DESIGN FOR SMALL ECCENTRICITY

$$\phi \beta P_o \geq P_u \qquad 52.7$$

$$P_o = 0.85 f'_c (A_g - A_{st}) + f_y A_{st} \qquad 52.8$$

$$P_u = 1.4 \ (\text{dead load axial force})$$
$$\quad + 1.7 \ (\text{live load axial force}) \qquad 52.9$$

$$0.85 f'_c (A_g - A_{st}) + A_{st} f_y = \frac{P_u}{\phi \beta} \qquad 52.10$$

$$A_g \left(0.85 f'_c (1 - \rho_g) + \rho_g f_y\right) = \frac{P_u}{\phi \beta} \qquad 52.11$$

$$\rho_g = \frac{A_{st}}{A_g} \qquad 52.12$$

CERM Chapter 53
Reinforced Concrete: Long Columns

Chapter, section, equation, figure, and table numbers correspond to CERM. For additional study material, go to the corresponding chapter and section number in CERM.

2. BRACED AND UNBRACED COLUMNS

$$Q = \frac{\sum P_u \Delta_o}{V_u l_c} \qquad 53.1$$

3. EFFECTIVE LENGTH

$$\text{effective length} = k l_u \qquad 53.2$$

$$\Psi = \frac{\sum_{\text{columns}} \dfrac{EI}{l_c}}{\sum_{\text{beams}} \dfrac{EI}{l}} \qquad 53.3$$

6. BUCKLING LOAD

$$P_c = \frac{\pi^2 EI}{(k l_u)^2} \qquad 53.6$$

7. COLUMNS IN BRACED STRUCTURES (NON-SWAY FRAMES)

$$M_c = \delta_{ns} M_2 \qquad 53.7$$

$$\delta_{ns} = \frac{C_m}{1 - \dfrac{P_u}{0.75 P_c}} \geq 1.0 \qquad 53.8$$

$$C_m = 0.6 + 0.4 \left(\frac{M_1}{M_2}\right) \geq 0.4 \qquad 53.9$$

$$M_{2,\min} = P_u (0.6 + 0.03h) \quad [h \text{ in inches}] \qquad 53.10$$

8. COLUMNS IN UNBRACED STRUCTURES (SWAY FRAMES)

$$M_1 = M_{1,ns} + \delta_s M_{1s} \qquad 53.11$$

$$M_2 = M_{2,ns} + \delta_s M_{2s} \qquad 53.12$$

$$\delta_s = \frac{1}{1 - \dfrac{\sum P_u}{0.75 \sum P_c}} \geq 1.0 \qquad 53.13$$

$$\delta_s = \frac{1}{1 - Q} \geq 1 \qquad 53.14$$

$$\frac{l_u}{r} \geq \frac{35}{\sqrt{\dfrac{P_u}{f'_c A_g}}} \qquad 53.15$$

CERM Chapter 54
Reinforced Concrete: Walls and Retaining Walls

Chapter, section, equation, figure, and table numbers correspond to CERM. For additional study material, go to the corresponding chapter and section number in CERM.

2. BEARING WALLS: EMPIRICAL METHOD

$$P_u \leq \phi P_{n,w} \leq 0.55\phi f_c' A_g \left(1 - \left(\frac{kl_c}{32h}\right)^2\right) \qquad 54.1$$

5. DESIGN OF RETAINING WALLS

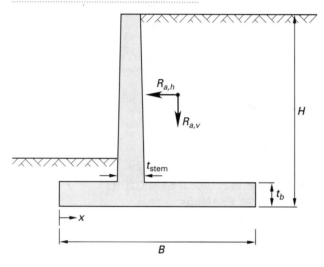

Figure 54.1 *Retaining Wall Dimensions*

$$\rho \approx \tfrac{1}{2}\rho_{\max} = 0.375\rho_{\text{balanced}} \qquad 54.2$$

$$R_{u,\text{trial}} = \rho f_y \left(1 - \left(\frac{\rho f_y}{(2)(0.85)f_c'}\right)\right) \qquad 54.3$$

$$M_{u,\text{stem}} = 1.7 R_{a,h} y_a \qquad 54.4$$

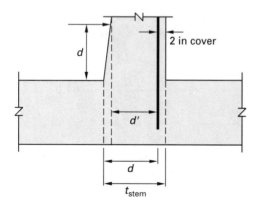

Figure 54.2 *Base of Stem Details*

$$d = \sqrt{\frac{M_{u,\text{stem}}}{\phi R_u w}} \qquad 54.5$$

$$t_{\text{stem}} = d + \text{cover} + \tfrac{1}{2}d_b \qquad 54.6$$

$$V_{u,\text{stem}} = 1.7 V_{\text{active}} + 1.7 V_{\text{surcharge}} \qquad 54.7$$

$$\phi V_n = 2\phi w d' \sqrt{f_c'} \qquad 54.8$$

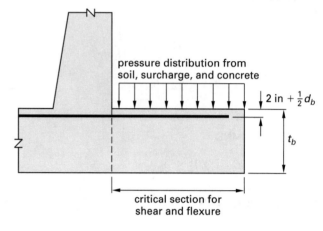

Figure 54.3 *Heel Details*

$$V_{u,\text{base}} = 1.4 V_{\text{soil}} + 1.4 V_{\text{heel weight}} \\ + 1.7 V_{\text{surcharge}} \qquad 54.9$$

$$M_{u,\text{base}} = 1.4 M_{\text{soil}} + 1.4 M_{\text{heel weight}} \\ + 1.7 M_{\text{surcharge}} \qquad 54.10$$

$$V_{u,\text{toe}} = 1.7 V_{\text{toe pressure}} - 1.4 V_{\text{toe weight}} \qquad 54.11$$

$$M_{u,\text{toe}} = 1.7 M_{\text{toe pressure}} - 1.4 M_{\text{toe weight}} \qquad 54.12$$

$$M_n = \frac{M_u}{\phi} = \rho f_y \left(1 - \frac{\rho f_y}{(2)(0.85)f_c'}\right) bd^2 \qquad 54.13$$

CERM Chapter 55
Reinforced Concrete: Footings

Chapter, section, equation, figure, and table numbers correspond to CERM. For additional study material, go to the corresponding chapter and section number in CERM.

1. INTRODUCTION

$$q_s = \frac{P_s}{A_f} + \gamma_c h + \gamma_s (H - h) \pm \frac{M_s\left(\dfrac{B}{2}\right)}{I_f} \leq q_a \qquad 55.1$$

$$I_f = \tfrac{1}{12}LB^3 \qquad 55.2$$

2. WALL FOOTINGS

$$q_u = \frac{P_u}{B} \qquad \text{55.3}$$

$$P_u = 1.4P_d + 1.7P_l \qquad \text{55.4}$$

$$\phi v_c \geq v_u \qquad \text{55.5}$$

$$v_c = 2\sqrt{f_c'} \qquad \text{55.6}$$

$$v_u = \left(\frac{q_u}{d}\right)\left(\frac{B-t}{2} - d\right) \qquad \text{55.7}$$

$$d = \frac{q_u(B-t)}{2(q_u + \phi v_c)} \qquad \text{55.8}$$

$$h = d + \tfrac{1}{2}(\text{diameter of } x \text{ bars}) \\ + \text{diameter of } z \text{ bars} + \text{cover} \\ \left[\begin{array}{c}\text{orthogonal reinforcement}\\ \text{under main steel}\end{array}\right] \qquad \text{55.9(a)}$$

$$h = d + \tfrac{1}{2}(\text{diameter of } x \text{ bars}) + \text{cover} \\ \left[\begin{array}{c}\text{orthogonal reinforcement}\\ \text{over main steel}\end{array}\right] \qquad \text{55.9(b)}$$

$$h \geq 6 \text{ in} + \text{diameter of } x \text{ bars} \\ + \text{diameter of } z \text{ bars} + 3 \text{ in} \qquad \text{55.10}$$

3. COLUMN FOOTINGS

$$v_u = \frac{q_u e}{d} \qquad \text{55.11}$$

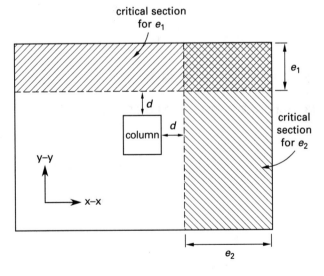

Figure 55.3 *Critical Section for One-Way Shear*

$$A_p = 2(b_1 + b_2)d \qquad \text{55.12}$$

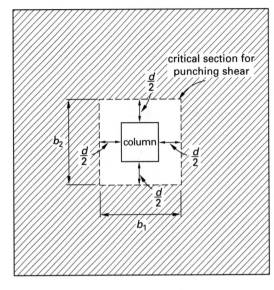

Figure 55.4 *Critical Section for Two-Way Shear*

$$v_u = \frac{P_u - R}{A_p} + \frac{\gamma_v M_u(0.5b_1)}{J} \qquad \text{55.13}$$

$$R = \frac{P_u b_1 b_2}{A_f} \qquad \text{55.14}$$

$$\gamma_v = 1 - \frac{1}{1 + \frac{2}{3}\sqrt{\frac{b_1}{b_2}}} \qquad \text{55.15}$$

$$J = \left(\frac{db_1^3}{6}\right)\left(1 + \left(\frac{d}{b_1}\right)^2 + 3\left(\frac{b_2}{b_1}\right)\right) \qquad \text{55.16}$$

$$v_c = (2 + y)\sqrt{f_c'} \qquad \text{55.17}$$

$$y = \min\{2, \ 4/\beta_c, \ 40d/b_o\} \qquad \text{55.18}$$

$$\beta_c = \frac{\text{column long side}}{\text{column short side}} \qquad \text{55.19}$$

$$b_o = \frac{A_p}{d} \qquad \text{55.20}$$

$$h = d + \left(\tfrac{1}{2}\right)(\text{diameter of } x \text{ bars} \\ + \text{diameter of } z \text{ bars}) + \text{cover} \qquad \text{55.21}$$

$$h \geq 6 \text{ in} + \text{diameter of } x \text{ bars} \\ + \text{diameter of } z \text{ bars} + 3 \text{ in} \qquad \text{55.22}$$

4. SELECTION OF FLEXURAL REINFORCEMENT

$$M_u = \frac{q_u L l^2}{2} \quad [\text{no column moment}] \qquad \text{55.23}$$

$$A_s = \frac{M_u}{\phi f_y(d - \lambda)} \qquad \text{55.24}$$

$$A_1 = A_{sd}\left(\frac{2}{\beta + 1}\right) \qquad 55.25$$

$$A_2 = A_{sd} - A_1 \qquad 55.26$$

5. DEVELOPMENT LENGTH OF FLEXURAL REINFORCEMENT

$$l_d = \frac{3d_b f_y \beta\gamma}{50\sqrt{f'_c}} \geq 12 \text{ in} \qquad 55.27$$

$$l_d = \frac{0.02\beta d_b f_y}{\sqrt{f'_c}} \geq 8d_b \geq 6 \text{ in} \qquad 55.28$$

6. TRANSFER OF FORCE AT COLUMN BASE

$$A_{db,\text{min}} = 0.005A_c \qquad 55.29$$

$$l_d = 0.02d_b\left(\frac{f_y}{\sqrt{f'_c}}\right)$$
$$\geq 0.0003d_b f_y \geq 8 \text{ in} \qquad 55.30$$

$$l_d \leq h - \text{cover} - \text{diameter of } x \text{ steel}$$
$$- \text{diameter of } z \text{ steel} \qquad 55.31$$

$$P_{\text{bearing,column}} = 0.85\phi f'_{c,\text{column}} A_c \qquad 55.32$$

$$P_{\text{bearing,footing}} = 0.85\alpha\phi f'_{c,\text{footing}} A_c \qquad 55.33$$

$$\alpha = \sqrt{\frac{A_{ff}}{A_c}} \leq 2 \qquad 55.34$$

$$A_{db} = \frac{P_u - \text{smaller of} \begin{Bmatrix} P_{\text{bearing,column}} \\ P_{\text{bearing,footing}} \end{Bmatrix}}{\phi f_y}$$
$$\geq A_{db,\text{min}} \qquad 55.35$$

CERM Chapter 56
Pretensioned Concrete

> Chapter, section, equation, figure, and table numbers correspond to CERM. For additional study material, go to the corresponding chapter and section number in CERM.

7. CREEP AND SHRINKAGE

$$\Delta_{\text{long term}} = C_c \Delta_{\text{immediate}} \qquad 56.1$$

8. PRESTRESS LOSSES

$$n = \frac{E_s}{E_c} \qquad 56.2$$

9. DEFLECTIONS

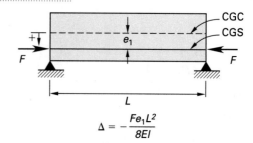

$$\Delta = -\frac{Fe_1 L^2}{8EI}$$

(a) straight tendons

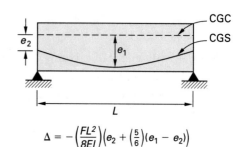

$$\Delta = -\left(\frac{FL^2}{8EI}\right)\left(e_2 + \left(\frac{5}{6}\right)(e_1 - e_2)\right)$$

(b) parabolically draped tendons

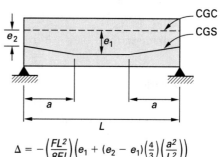

$$\Delta = -\left(\frac{FL^2}{8EI}\right)\left(e_1 + (e_2 - e_1)\left(\frac{4}{3}\right)\left(\frac{a^2}{L^2}\right)\right)$$

(c) shaped tendons

[a]CGC—centroid of concrete; CGS—centroid of prestressing tendons

Figure 56.2 *Midspan Deflections from Prestressing*[a]

13. ACI CODE PROVISIONS FOR STRENGTH

$$f_{\text{ps}} = f_{\text{pu}}\left(1 - \left(\frac{\gamma_p}{\beta_1}\right)\left(\rho_p\left(\frac{f_{\text{pu}}}{f'_c}\right) + \left(\frac{d}{d_p}\right)(\omega - \omega')\right)\right)$$
$$\text{[bonded tendons]} \qquad 56.3$$

$$\rho_p = \frac{A_{\text{sp}}}{bd_p} \qquad 56.4$$

$$\rho_p\left(\frac{f_{\text{pu}}}{f'_c}\right) + \left(\frac{d}{d_p}\right)(\omega - \omega') \geq 0.17 \qquad 56.5$$

14. ANALYSIS OF PRESTRESSED BEAMS

$$f_c = \frac{-P}{A} + \frac{Pey}{I} - \frac{M_s y}{I} \qquad 56.6$$

15. SHEAR IN PRESTRESSED SECTIONS

$$V_u < \phi V_n \qquad 56.7$$

$$V_n = V_c + V_s \qquad 56.8$$

CERM Chapter 57
Composite Concrete and Steel Bridge Girders

Chapter, section, equation, figure, and table numbers correspond to CERM. For additional study material, go to the corresponding chapter and section number in CERM.

5. EFFECTIVE SLAB WIDTH

$$b_e = \text{smaller of } \{L/4; \ b_o\} \qquad 57.1$$

$$b_e = \text{smallest of } \{L/4; \ 12t; \ b_o\} \qquad 57.2$$

$$b_e = \text{smallest of } \{L/12; \ 6t; \ \tfrac{1}{2}b_o\} \qquad 57.3$$

6. SECTION PROPERTIES

$$n = \frac{E_s}{E_c} \qquad 57.4$$

CERM Chapter 59
Structural Steel: Beams

Chapter, section, equation, figure, and table numbers correspond to CERM. For additional study material, go to the corresponding chapter and section number in CERM.

1. TYPES OF BEAMS

$$\frac{h}{t_w} \le \frac{970}{\sqrt{F_y}} \quad \text{[beam criterion]} \qquad 59.1$$

3. BENDING STRESS IN STEEL BEAMS

$$f_b = \frac{Mc}{I} = \frac{M}{S} \qquad 59.2$$

4. COMPACT SECTIONS

$$\frac{b_f}{2t_f} \le \frac{65.0}{\sqrt{F_y}} \qquad 59.3$$

$$\frac{d}{t_w} \le \frac{640}{\sqrt{F_y}} \quad \begin{bmatrix} \text{webs in flexural} \\ \text{compression only} \end{bmatrix} \qquad 59.4$$

5. LATERAL BRACING

$$L_c = \text{smaller of } \left\{ \begin{array}{c} \dfrac{76b_f}{\sqrt{F_y}} \\[2ex] \dfrac{20{,}000}{\left(\dfrac{d}{A_f}\right)F_y} \end{array} \right\} \quad \text{[F1-2]} \quad 59.5$$

$$L_u = \text{larger of } \left\{ \begin{array}{c} \dfrac{20{,}000C_b}{\left(\dfrac{d}{A_f}\right)F_y} \\[3ex] r_T\sqrt{\dfrac{102{,}000C_b}{F_y}} \end{array} \right\} \quad \text{[F1-6]} \quad 59.6$$

6. ALLOWABLE BENDING STRESS: W SHAPES BENDING ABOUT MAJOR AXIS

$$F_b = \left(\frac{2}{3} - \frac{F_y\left(\dfrac{l}{r_T}\right)^2}{1530 \times 10^3 C_b} \right) F_y \qquad 59.7$$

$$F_b = \frac{170 \times 10^3 C_b}{\left(\dfrac{l}{r_T}\right)^2} \le 0.60 F_y \qquad 59.8$$

$$F_b = \frac{12 \times 10^3 C_b}{\dfrac{ld}{A_f}} \le 0.60 F_y \qquad 59.9$$

9. SHEAR STRESS IN STEEL BEAMS

$$f_v = \frac{V}{A_w} = \frac{V}{dt_w} \qquad 59.10$$

10. CONCENTRATED WEB FORCES

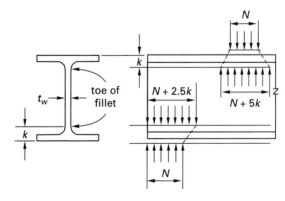

Figure 59.7 *Nomenclature for Web Yielding Calculations*

$$N_{min} = \frac{R - 2R_1}{R_2} \quad \text{[interior]} \qquad 59.11$$

$$N_{min} = \frac{R - R_1}{R_2} \quad \text{[ends]} \qquad 59.12$$

$$R_1 = 0.66 F_{yw} t_w (2.5k) \quad [F_{yw} \text{ in kips/in}^2] \quad 59.13$$

$$R_2 = 0.66 F_{yw} t_w \quad [F_{yw} \text{ in kips/in}^2] \qquad 59.14$$

$$N_{min} = \frac{R - 2R_3}{R_4} \quad \text{[interior]} \qquad 59.15$$

$$N_{min} = \frac{R - R_3}{R_4} \quad \text{[ends]} \qquad 59.16$$

$$R_3 = 34 t_w^2 \sqrt{\frac{F_{yw} t_f}{t_w}} \quad [F_{yw} \text{ in kips/in}^2] \quad 59.17$$

$$R_4 = 34 t_w^2 \left(\frac{3}{d}\right) \left(\frac{t_w}{t_f}\right)^{1.5} \sqrt{\frac{F_{yw} t_f}{t_w}}$$
$$[F_{yw} \text{ in kips/in}^2] \quad 59.18$$

14. PLASTIC DESIGN OF CONTINUOUS BEAMS

$$\text{factored load} = 1.7 \times \text{dead load}$$
$$+ 1.7 \times \text{live load} \qquad 59.19$$

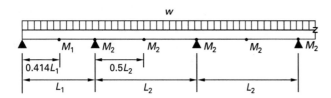

Figure 59.13 *Ultimate Moments on a Uniformly Loaded Continuous Beam (plastic hinges shown as •)*

$$Z_x = \frac{M_p}{F_y} \qquad 59.20$$

$$V_{\text{factored,max}} \le 0.55 F_y t_w d \qquad 59.21$$

$$\frac{l_{cr}}{r_y} = \frac{1375}{F_y} + 25 \quad \left[+1.0 > \frac{M}{M_p} > -0.5\right] \qquad 59.22$$

$$\frac{l_{cr}}{r_y} = \frac{1375}{F_y} \quad \left[-0.5 \ge \frac{M}{M_p} > -1.0\right] \qquad 59.23$$

15. ULTIMATE PLASTIC MOMENTS

$$M_1 = \left(\tfrac{3}{2} - \sqrt{2}\right) w L_1^2$$
$$\approx 0.0858 w L_1^2 \qquad 59.24$$

$$M_2 = \frac{wL^2}{16} \qquad 59.25$$

16. ULTIMATE SHEARS

$$V_{max} = \frac{wL}{2} + \frac{M_{max}}{L} = 0.5858 wL \qquad 59.26$$

18. UNSYMMETRICAL BENDING

$$\frac{f_a}{F_a} + \frac{f_{bx}}{F_{bx}} + \frac{f_{by}}{F_{by}} \le 1.0 \qquad 59.27$$

$$\frac{f_{bx}}{F_{bx}} + \frac{f_{by}}{F_{by}} \le 1.0 \qquad 59.28$$

21. BEAM BEARING PLATES

$$A_1 = \frac{R}{0.35 f'_c} \qquad 59.29$$

$$A_1 = \left(\frac{1}{A_2}\right) \left(\frac{R}{0.35 f'_c}\right)^2 \qquad 59.30$$

$$N = \frac{R - R_1}{R_2} \qquad 59.31$$

$$N = \frac{R - R_3}{R_4} \qquad 59.32$$

$$B = \frac{A_1}{N} \qquad 59.33$$

$$f_p = \frac{R}{BN} \qquad 59.34$$

$$n = \frac{B}{2} - k \qquad 59.35$$

$$t = \sqrt{\frac{3 f_p n^2}{F_b}} \qquad 59.36$$

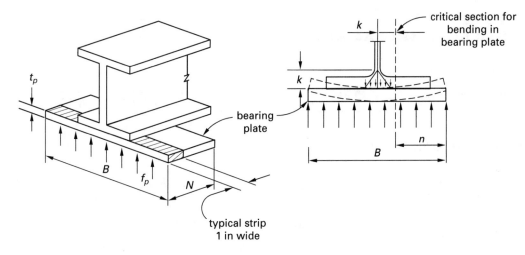

Figure 59.15 *Nomenclature for Beam Bearing Plate*

CERM Chapter 60
Structural Steel: Tension Members

Chapter, section, equation, figure, and table numbers correspond to CERM. For additional study material, go to the corresponding chapter and section number in CERM.

2. AXIAL TENSILE STRESS

$$f_t = \frac{P}{A} \qquad\qquad 60.1$$

3. GROSS AREA

$$A_g = bt \qquad\qquad 60.2$$

4. NET AREA

$$b_n = b - \sum d_h + \sum \frac{s^2}{4g} \qquad\qquad 60.3$$

$$A_n = b_n t \qquad\qquad 60.4$$

$$A_n = A_g - \sum d_h t + \left(\frac{\sum s^2}{4g}\right) t \qquad\qquad 60.5$$

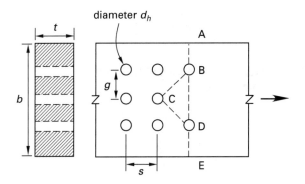

Figure 60.4 *Tension Member with Uniform Thickness and Unstaggered Holes*

5. EFFECTIVE NET AREA

$$A_e = U A_n \qquad\qquad 60.6$$

$$A_e = U A_g \qquad\qquad 60.7$$

7. BLOCK SHEAR STRENGTH

$$P_t = A_v F_v + A_t F_t \qquad\qquad 60.8$$

$$F_v = 0.30 F_u \qquad\qquad 60.9$$

$$F_t = 0.50 F_u \qquad\qquad 60.10$$

9. ANALYSIS OF TENSION MEMBERS

$$P_t = 0.60 F_y A_g \quad \text{[yielding]} \qquad\qquad 60.11$$

$$P_t = 0.50 F_u A_e \quad \text{[fracture]} \qquad\qquad 60.12$$

$$P_t = 0.30 F_u A_v + 0.50 F_u A_t \quad \text{[block shear]} \qquad 60.13$$

CERM Chapter 61
Structural Steel: Compression Members

Chapter, section, equation, figure, and table numbers correspond to CERM. For additional study material, go to the corresponding chapter and section number in CERM.

2. EULER'S COLUMN BUCKLING THEORY

$$P_e = \frac{\pi^2 EI}{L^2} \qquad 61.1$$

$$F_e = \frac{P_e}{A} = \frac{\pi^2 E}{\left(\frac{L}{r}\right)^2} \qquad 61.2$$

3. EFFECTIVE LENGTH

$$F_e = \frac{\pi^2 E}{\left(\frac{KL}{r}\right)^2} \qquad 61.3$$

$$G = \frac{\sum\left(\frac{I}{L}\right)_{\text{columns}}}{\sum\left(\frac{I}{L}\right)_{\text{beams}}} \qquad 61.4$$

5. SLENDERNESS RATIO

$$\text{SR} = \frac{KL}{r} \qquad 61.5$$

$$C_c = \sqrt{\frac{2\pi^2 E}{F_y}} \qquad 61.6$$

7. ALLOWABLE COMPRESSIVE STRESS

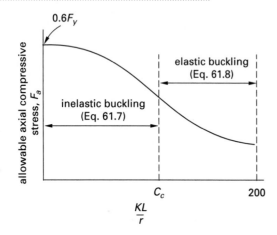

Figure 61.4 *Allowable Compressive Stress Versus Slenderness Ratio*

$$F_a = \frac{\left(1 - \frac{\left(\frac{KL}{r}\right)^2}{2C_c^2}\right) F_y}{\frac{5}{3} + \frac{3\left(\frac{KL}{r}\right)}{8C_c} - \frac{\left(\frac{KL}{r}\right)^3}{8C_c^3}} \qquad \left[\frac{KL}{r} \leq C_c\right] \quad 61.7$$

$$F_a = \frac{12\pi^2 E}{23\left(\frac{KL}{r}\right)^2} \qquad \left[\frac{KL}{r} > C_c\right] \quad 61.8$$

$$F_a = C_a F_y \qquad 61.9$$

8. ANALYSIS OF COLUMNS

$$\text{SR} = \text{larger of} \begin{cases} \dfrac{K_x L_x}{r_x} \\[2ex] \dfrac{K_y L_y}{r_y} \end{cases} \qquad 61.10$$

$$P_{\max} = F_a A \qquad 61.11$$

9. DESIGN OF COLUMNS

$$\text{effective length} = K_y L_y \qquad 61.12$$

$$K_x L'_x = \frac{K_x L_x}{\frac{r_x}{r_y}} \qquad 61.13$$

10. LOCAL BUCKLING

$$\frac{b}{t} \leq \frac{H}{\sqrt{F_y}} \qquad \text{[plate element]} \qquad 61.14$$

$$\frac{D}{t} \leq \frac{3300}{F_y} \qquad \text{[circular tubes]} \qquad 61.15$$

$$F'_a = Q_s F_a \qquad 61.16$$

11. MISCELLANEOUS COMBINATIONS IN COMPRESSION

$$L_b = \frac{\text{SR}_{\max} r_z}{K} \qquad \text{[double angles]} \qquad 61.17$$

12. COLUMN BASE PLATES

$$A_{\text{plate}} = \frac{\text{column load}}{F_p} \qquad 61.18$$

$$f_p = \frac{\text{column load}}{\text{actual plate area}} \qquad 61.19$$

$$A_1 = \frac{\text{column load}}{F_p} \qquad 61.20$$

$$N \approx \sqrt{A_1} + \tfrac{1}{2}(0.95d - 0.80b_f) \qquad 61.21$$

$$B = \frac{A_1}{N} \qquad 61.22$$

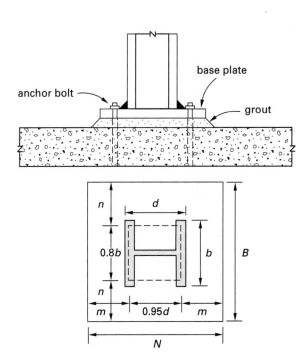

Figure 61.7 *Column Base Plate*

$$t = \text{larger} \begin{Bmatrix} m \\ n \end{Bmatrix} \sqrt{\frac{3f_p}{F_b}}$$

$$= \text{larger} \begin{Bmatrix} 2m \\ 2n \end{Bmatrix} \sqrt{\frac{f_p}{F_y}} \qquad 61.23$$

$$t = L\sqrt{\frac{3f_p}{F_b}} \qquad 61.24$$

$$F_p = \frac{P}{A} = \frac{P}{2L(d + b - 2L)} \qquad 61.25$$

CERM Chapter 62
Structural Steel: Beam-Columns

Chapter, section, equation, figure, and table numbers correspond to CERM. For additional study material, go to the corresponding chapter and section number in CERM.

1. INTRODUCTION

$$f_{\max} = f_a + (\text{AF})_x f_{bx} + (\text{AF})_y f_{by} \qquad 62.1$$

$$\frac{f_a}{F_a} + (\text{AF})_x \left(\frac{f_{bx}}{F_{bx}}\right) + (\text{AF})_y \left(\frac{f_{by}}{F_{by}}\right) = 1.0 \qquad 62.2$$

2. SMALL AXIAL COMPRESSION

$$\frac{f_a}{F_a} + \frac{f_{bx}}{F_{bx}} + \frac{f_{by}}{F_{by}} \le 1.0 \qquad 62.3$$

3. LARGE AXIAL COMPRESSION

$$\frac{f_a}{F_a} + \frac{C_{mx}f_{bx}}{\left(1 - \dfrac{f_a}{F'_{ex}}\right)F_{bx}} + \frac{C_{my}f_{by}}{\left(1 - \dfrac{f_a}{F'_{ey}}\right)F_{by}} \le 1.0 \qquad 62.4$$

$$F'_e = \frac{12\pi^2 E}{23\left(\dfrac{KL_b}{r_b}\right)^2} \qquad 62.5$$

$$C_m = 0.6 - 0.4\left(\frac{M_1}{M_2}\right) \begin{bmatrix} \text{braced against} \\ \text{joint translation} \end{bmatrix} \qquad 62.6$$

$$\frac{f_a}{0.60F_y} + \frac{f_{bx}}{F_{bx}} + \frac{f_{by}}{F_{by}} \le 1.0 \qquad 62.7$$

CERM Chapter 63
Structural Steel: Plate Girders

Chapter, section, equation, figure, and table numbers correspond to CERM. For additional study material, go to the corresponding chapter and section number in CERM.

1. INTRODUCTION

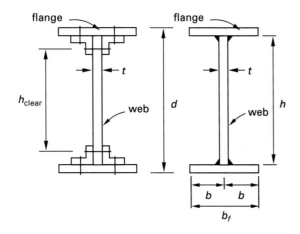

Figure 63.1 Elements of a Plate Girder

2. DEPTH-THICKNESS RATIOS

$$\frac{h}{t_w} \leq \frac{2000}{\sqrt{F_y}} \quad [\text{spacing} \leq 1.5d] \qquad 63.1$$

$$\frac{h}{t_w} \leq \frac{14{,}000}{\sqrt{F_{yf}(F_{yf} + 16.5)}} \quad [\text{spacing} > 1.5d] \qquad 63.2$$

3. SHEAR STRESS

$$f_v = \frac{V_{\max}}{ht_w} \qquad 63.3$$

If $\dfrac{h}{t_w} > \dfrac{380}{\sqrt{f_y}}$,

$$F_v = \frac{C_v F_y}{2.89} \leq 0.40 F_y \qquad 63.4$$

$$C_v = \frac{45{,}000 k_v}{F_y \left(\dfrac{h}{t_w}\right)^2} \quad [C_v < 0.8] \qquad 63.5$$

$$C_v = \left(\frac{190}{\dfrac{h}{t_w}}\right)\sqrt{\frac{k_v}{F_y}} \quad [C_v > 0.8] \qquad 63.6$$

$$k_v = 4.00 + \frac{5.34}{\left(\dfrac{a}{h}\right)^2} \quad \left[\frac{a}{h} < 1.0\right] \qquad 63.7$$

$$k_v = 5.34 + \frac{4.00}{\left(\dfrac{a}{h}\right)^2} \quad \left[\frac{a}{h} > 1.0\right] \qquad 63.8$$

$$F_v = \frac{152\sqrt{F_y}}{\dfrac{h}{t}} \quad \left[\frac{h}{t} < \frac{548}{\sqrt{F_y}}\right] \qquad 63.9$$

$$F_v = \frac{83{,}150}{\left(\dfrac{h}{t}\right)^2} \quad \left[\frac{h}{t} > \frac{548}{\sqrt{F_y}}\right] \qquad 63.10$$

5. DESIGN OF GIRDER FLANGES

$$\frac{h}{t} > \frac{970}{\sqrt{F_b}} \qquad 63.11$$

$$A_f = \frac{M_x}{F_b h} - \frac{th}{6} \qquad 63.12$$

$$A_f = b_f t_f \qquad 63.13$$

6. WIDTH-THICKNESS RATIOS

$$\frac{b}{t_f} < \frac{95}{\sqrt{\dfrac{F_{yf}}{k_c}}} \qquad 63.14$$

$$b = \frac{b_f}{2} \qquad 63.15$$

$$k_c = \begin{cases} \dfrac{4.05}{\left(\dfrac{h}{t}\right)^{0.46}} & \dfrac{h}{t} > 70 \\[3mm] 1.0 & \dfrac{h}{t} \leq 70 \end{cases}$$

7. REDUCTION IN FLANGE STRESS

$$F_b' \leq F_b\left(1.0 - 0.0005\left(\frac{A_w}{A_f}\right)\left(\frac{h}{t} - \frac{760}{\sqrt{F_b}}\right)\right) \qquad 63.16$$

$$A_w = ht \qquad 63.17$$

$$A_f = b_f t_f \qquad 63.18$$

9. LOCATION OF INTERIOR STIFFENERS

$$\frac{a}{h} \leq \left(\frac{260}{\left(\dfrac{h}{t_w}\right)}\right)^2 \quad \text{and } 3.0 \qquad 63.19$$

$$F_v = \frac{F_y}{2.89}\left(C_v + \frac{1 - C_v}{1.15\sqrt{1 + \left(\dfrac{a}{h}\right)^2}}\right)$$

$$\leq 0.40 F_y \qquad 63.20$$

10. MAXIMUM BENDING STRESS

$$f_b \leq \left(0.825 - 0.375\left(\frac{f_v}{F_v}\right)\right)F_y < 0.60F_y \qquad 63.21$$

11. DESIGN OF INTERMEDIATE STIFFENERS

$$A_{st} = \left(\frac{1 - C_v}{2}\right)\left(\frac{a}{h} - \frac{\left(\frac{a}{h}\right)^2}{\sqrt{1 + \left(\frac{a}{h}\right)^2}}\right)\left(\frac{F_{y,web}}{F_{y,stiffener}}\right)Dht$$

$$63.22$$

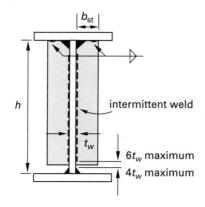

Figure 63.2 Intermediate Stiffeners

$$A'_{st} = \left(\frac{f_v}{F_v}\right)A_{st} \qquad 63.23$$

$$I_{st} = \frac{t_{st}b_{st}^3}{12} \qquad 63.24$$

$$I_{st} \geq \left(\frac{h}{50}\right)^4 \qquad 63.25$$

12. DESIGN OF BEARING STIFFENERS

$$\frac{b_{st}}{t_{st}} \leq \frac{95}{\sqrt{F_y}} \qquad 63.26$$

$$\frac{l}{r} = \frac{0.75h}{0.25(2b_{st} + t_w)} \qquad 63.27$$

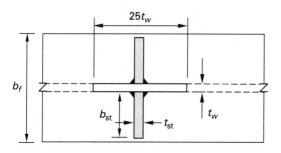

Figure 63.3 Bearing Stiffener (top view)

$$t_{st} = \frac{\dfrac{load}{F_a} - 25t_w^2}{2b_{st}} \qquad 63.28$$

$$t_{st} = \frac{\dfrac{load}{0.60F_y}}{2b_{st}} \qquad 63.29$$

CERM Chapter 64
Structural Steel: Composite Steel Members for Buildings

> Chapter, section, equation, figure, and table numbers correspond to CERM. For additional study material, go to the corresponding chapter and section number in CERM.

2. EFFECTIVE WIDTH OF CONCRETE SLAB

$$b = \text{smaller of} \left\{\begin{array}{c} \dfrac{L}{4} \\ s \end{array}\right\} \qquad \text{[interior beams]} \qquad 64.1$$

3. SECTION PROPERTIES

$$n = \frac{E}{E_c} \qquad 64.2$$

$$\overline{y}_b = \frac{A_s\left(\dfrac{d}{2}\right) + A_{ctr}(d + Y2)}{A_s + A_{ctr}} \qquad 64.3$$

$$A_{ctr} = \left(\frac{b}{n}\right)t \qquad 64.4$$

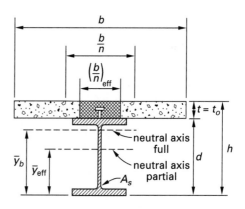

[a]This figure depicts the neutral axis as being located within the steel beam, which is usually the case.

Figure 64.2 Nomenclature for Composite Beam with Solid Slab[a]

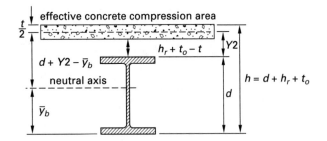

[a]This figure depicts the neutral axis as being located within the steel beam, which is usually the case.

Figure 64.3 *Nomenclature for Fully Composite Beam with Steel Deck[a]*

$$I_{\text{tr}} = I_s + A_s \left(\bar{y}_b - \frac{d}{2} \right)^2 + \left(\frac{1}{12} \right) \left(\frac{b}{n} \right) t^3$$
$$+ A_{\text{ctr}}(d + Y2 - \bar{y}_b)^2 \qquad \textbf{64.5}$$

$$S_{\text{tr}} = \frac{I_{\text{tr}}}{\bar{y}_b} \qquad \textbf{64.6}$$

$$Y2 = \frac{t}{2} \quad \text{[with solid slab]} \qquad \textbf{64.7}$$

$$Y2 = h_r + \frac{t_o}{2} \quad \text{[with slab on metal deck]} \qquad \textbf{64.8}$$

$$S_t = \frac{I_{\text{tr}}}{h - \bar{y}_b} \qquad \textbf{64.9}$$

$$\left(\frac{b}{n} \right)_{\text{eff}} = \left(\frac{A_s}{t} \right) \left(\frac{\bar{y}_{\text{eff}} - \frac{d}{2}}{d + \frac{t}{2} - \bar{y}_{\text{eff}}} \right) \leq \frac{b}{n}$$
$$\text{[solid slab]} \qquad \textbf{64.10}$$

$$\left(\frac{b}{n} \right)_{\text{eff}} = \left(\frac{A_s}{t} \right) \left(\frac{\bar{y}_{\text{eff}} - \frac{d}{2}}{d + Y2 - \bar{y}_{\text{eff}}} \right) \leq \frac{b}{n}$$
$$\text{[slab on steel deck]} \qquad \textbf{64.11}$$

$$\bar{y}_{\text{eff}} = \frac{I_{\text{eff}}}{S_{\text{eff}}} \qquad \textbf{64.12}$$

$$S_{\text{eff}} = S_s + \sqrt{\frac{V_h'}{V_h}} (S_{\text{tr}} - S_s) \qquad \textbf{64.13}$$

$$V_h = \frac{0.85 f_c' A_c}{2} \qquad \textbf{64.14}$$

$$V_h = \frac{F_y A_s}{2} \qquad \textbf{64.15}$$

$$V_h' = q N_1 \qquad \textbf{64.16}$$

$$I_{\text{eff}} = I_s + \sqrt{\frac{V_h'}{V_h}} (I_{\text{tr}} - I_s) \qquad \textbf{64.17}$$

$$S_{t,\text{eff}} = \frac{I_{\text{eff}}}{h - \bar{y}_{\text{eff}}} \qquad \textbf{64.18}$$

4. WORKING STRESSES WITH SHORING

- For full composite action,

$$f_s = \frac{M_D + M_L}{S_{\text{tr}}} \qquad \textbf{64.19}$$

$$f_c = \frac{M_D + M_L}{n S_t} \qquad \textbf{64.20}$$

- For partial composite action,

$$f_s = \frac{M_D + M_L}{S_{\text{eff}}} \qquad \textbf{64.21}$$

$$f_c = \left(\frac{M_D + M_L}{S_{t,\text{eff}}} \right) \left(\frac{\left(\frac{b}{n} \right)_{\text{eff}}}{b} \right) \qquad \textbf{64.22}$$

5. WORKING STRESSES WITHOUT SHORING

$$f_s = \frac{M_D}{S_s} \qquad \textbf{64.23}$$

- For full composite action,

$$f_s = \frac{M_D}{S_s} + \frac{M_L}{S_{\text{tr}}} \qquad \textbf{64.24}$$

$$f_c = \frac{M_L}{n S_t} \qquad \textbf{64.25}$$

- For partial composite action,

$$f_s = \frac{M_D}{S_s} + \frac{M_L}{S_{\text{eff}}} \qquad \textbf{64.26}$$

$$f_c = \frac{M_L}{S_{t,\text{eff}}} \left(\frac{\left(\frac{b}{n} \right)_{\text{eff}}}{b} \right) \qquad \textbf{64.27}$$

7. SHEAR CONNECTORS

$$N_1 = \frac{V_h}{q} \quad \text{[full composite action]} \qquad \textbf{64.28}$$

$$N_1 = \frac{V_h'}{q} \quad \text{[partial composite action]} \qquad \textbf{64.29}$$

CERM Chapter 65
Structural Steel: Connectors

Chapter, section, equation, figure, and table numbers correspond to CERM. For additional study material, go to the corresponding chapter and section number in CERM.

2. HOLE SPACING AND EDGE DISTANCES

$$s \geq \frac{2P}{F_u t} + \frac{d}{2} \qquad 65.1$$

$$L_e \geq \frac{2P}{F_u t} \qquad 65.2$$

4. ALLOWABLE LOADS FOR FASTENERS

$$\text{reduction factor} = 1 - \frac{f_t A_b}{T_b} \qquad 65.3$$

$$T_{b,\min} = 0.70 A_{b,t} F_u \quad [\text{rounded}] \qquad 65.4$$

5. ALLOWABLE BEARING STRESS

$$F_p = 1.20 F_u \qquad 65.5$$

8. ULTIMATE STRENGTH ANALYSIS OF ECCENTRIC SHEAR CONNECTIONS

$$P = C A_b F_v \quad [C \text{ from } \textit{AISC Manual}] \qquad 65.6$$

10. COMBINED SHEAR AND TENSION CONNECTIONS

$$f_{t,\max} = \frac{Mc}{I} = \frac{Pec}{I} \qquad 65.7$$

$$T_{\max} = f_{t,\max} A_b \qquad 65.8$$

CERM Chapter 66
Structural Steel: Welding

Chapter, section, equation, figure, and table numbers correspond to CERM. For additional study material, go to the corresponding chapter and section number in CERM.

3. FILLET WELDS

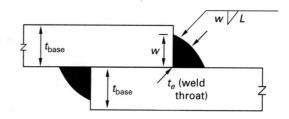

Figure 66.4 *Fillet Weld*

$$t_e = 0.707w \qquad 66.1$$

$$t_e = w \quad [w \leq \tfrac{3}{8} \text{ in}] \qquad 66.2$$

$$t_e = 0.707w + 0.11 \quad [w \geq \tfrac{3}{8} \text{ in}] \qquad 66.3$$

$$F_v = 0.30 F_{u,\text{rod}} \qquad 66.4$$

$$R_w = 0.30 t_e F_{u,\text{rod}} \qquad 66.5$$

4. CONCENTRIC TENSION CONNECTIONS

$$f_v = \frac{P}{A_{\text{weld}}} = \frac{P}{L_{\text{weld}} t_e} \qquad 66.6$$

5. NONCONCENTRIC TENSION CONNECTIONS

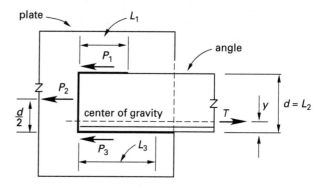

Figure 66.5 *Balanced Weld Group*

$$P_3 = T\left(1 - \frac{y}{d}\right) - \frac{P_2}{2} \qquad 66.7$$

$$P_2 = R_w L_2 = R_w d \qquad 66.8$$

$$P_1 = T - P_2 - P_3 \qquad \textit{66.9}$$

$$L_1 = \frac{P_1}{R_w} \qquad \textit{66.10}$$

$$L_3 = \frac{P_3}{R_w} \qquad \textit{66.11}$$

7. ELASTIC METHOD FOR ECCENTRIC SHEAR/TORSION CONNECTIONS

$$J = I_x + I_y \qquad \textit{66.12}$$

$$f_{v,t} = \frac{Mr}{J} = \frac{Per}{J} \qquad \textit{66.13}$$

$$f_{v,d} = \frac{P}{A} \qquad \textit{66.14}$$

$$f_v = \sqrt{(f_{v,t,y} + f_{v,d})^2 + (f_{v,t,x})^2} \qquad \textit{66.15}$$

8. AISC ULTIMATE STRENGTH METHOD FOR ECCENTRIC SHEAR/TORSION CONNECTIONS

$$P = CC_1 Dl \quad \begin{bmatrix} C \text{ and } C_1 \text{ from} \\ AISC \text{ Manual} \end{bmatrix} \qquad \textit{66.16}$$

9. CONNECTIONS WITH SHEAR AND BENDING STRESSES

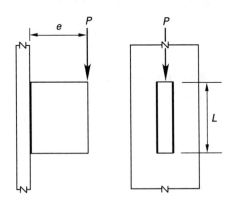

Figure 66.7 Welded Connection in Combined Shear and Bending

$$f_v = \frac{P}{A_{\text{weld}}} = \frac{P}{2Lt_e} \qquad \textit{66.17}$$

$$f_b = \frac{Mc}{I} = \frac{M}{S} = \frac{Pe}{S} \qquad \textit{66.18}$$

$$f = \sqrt{f_v^2 + f_b^2} \qquad \textit{66.19}$$

CERM Chapter 68
Properties of Masonry

> Chapter, section, equation, figure, and table numbers correspond to CERM. For additional study material, go to the corresponding chapter and section number in CERM.

6. MODULUS OF ELASTICITY

$$E_m = 700 f_m' \quad \text{[clay masonry]} \qquad \textit{68.1}$$

$$E_m = 900 f_m' \quad \text{[concrete masonry]} \qquad \textit{68.2}$$

17. DEVELOPMENT LENGTH

$$l_d = 0.0015 d_b F_s \qquad \textit{68.3(b)}$$

CERM Chapter 69
Masonry Walls

> Chapter, section, equation, figure, and table numbers correspond to CERM. For additional study material, go to the corresponding chapter and section number in CERM.

7. ASD WALL DESIGN: FLEXURE—REINFORCED

$$F_b = \tfrac{1}{3} f_m' \quad [2.3.3.2.2] \qquad \textit{69.1}$$

$$F_v = \sqrt{f_m'} \quad [2.3.5.2.2] \qquad \textit{69.2}$$

$$\rho = \frac{A_s}{bd} \qquad \textit{69.3}$$

$$k = \sqrt{2\rho n + (\rho n)^2} - \rho n \qquad \textit{69.4}$$

$$j = 1 - \frac{k}{3} \qquad \textit{69.5}$$

$$M_m = F_b bd^2 \left(\frac{jk}{2}\right) \qquad \textit{69.6}$$

$$M_s = A_s F_s jd \qquad \textit{69.7}$$

$$M_R = \text{the lesser of } M_m \text{ and } M_s \qquad \textit{69.8}$$

$$V_R = F_v bd \qquad \textit{69.9}$$

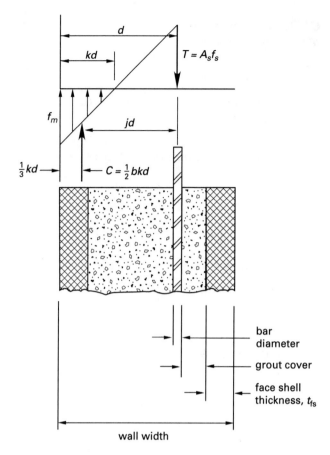

Figure 69.1 *Stress Distribution for Fully Grouted Masonry*

$$\rho_{\text{bal}} = \frac{nF_b}{2F_s\left(n + \dfrac{F_s}{F_b}\right)} \quad \text{[balanced]} \qquad 69.10$$

$$k = \frac{-A_s n - t_{\text{fs}}(b - b_w)}{db_w}$$

$$+ \frac{\sqrt{\begin{array}{c}(A_s n + t_{\text{fs}}(b - b_w))^2 \\ + t_{\text{fs}}^2 b_w(b - b_w) + 2db_w A_s n\end{array}}}{db_w}$$

$$69.11$$

$$j = \left(\frac{1}{kdb_w + t_{\text{fs}}(b - b_w)\left(2 - \dfrac{t_{\text{fs}}}{kd}\right)}\right)$$

$$\times \left(kb_w\left(d - \frac{kd}{3}\right) + \left(\frac{2t_{\text{fs}}(b - b_w)}{kd^2}\right)\right.$$

$$\left.\times \left((kd - t_{\text{fs}})\left(d - \frac{t_{\text{fs}}}{2}\right) + \left(\frac{t_{\text{fs}}}{2}\right)\left(d - \frac{t_{\text{fs}}}{3}\right)\right)\right)$$

$$69.12$$

$$M_m = \tfrac{1}{2}F_b kdb_w\left(d - \frac{kd}{3}\right) + F_b t_{\text{fs}}(b - b_w)$$

$$\times \left(\left(1 - \frac{t_{\text{fs}}}{kd}\right)\left(d - \frac{t_{\text{fs}}}{2}\right) + \left(\frac{t_{\text{fs}}}{2kd}\right)\left(d - \frac{t_{\text{fs}}}{3}\right)\right)$$

$$69.13$$

$$M_s = A_s F_s jd \qquad 69.14$$

$$M_R = \text{the lesser of } M_m \text{ and } M_s \qquad 69.15$$

$$V_R = F_v\left(bt_{\text{fs}} + b_w(d - t_{\text{fs}})\right) \qquad 69.16$$

8. ASD WALL DESIGN: AXIAL COMPRESSION AND FLEXURE—UNREINFORCED

$$\frac{f_a}{F_a} + \frac{f_b}{F_b} \le 1 \qquad 69.17$$

$$P \le \tfrac{1}{4}P_e \qquad 69.18$$

$$P_e = \left(\frac{\pi^2 E_m I}{h^2}\right)\left(1 - 0.577\left(\frac{e}{r}\right)\right)^3 \qquad 69.19$$

$$F_a = \tfrac{1}{4}f_m'\left(1 - \left(\frac{h}{140r}\right)^2\right) \quad [h/r \le 99] \quad 69.20$$

$$F_a = \tfrac{1}{4}f_m'\left(\frac{70r}{h}\right)^2 \quad [\text{with } h/r > 99] \qquad 69.21$$

$$F_b = \tfrac{1}{3}f_m' \qquad 69.22$$

$$R = 1 - \left(\frac{h}{140r}\right)^2 \quad [h/r \le 99] \qquad 69.23$$

$$R = \left(\frac{70r}{h}\right)^2 \quad [h/r > 99] \qquad 69.24$$

9. ASD WALL DESIGN: AXIAL COMPRESSION AND FLEXURE—REINFORCED

$$F_b = \tfrac{1}{3}f_m' \quad [2.3.3.2.2] \qquad 69.25$$

$$P_a = (0.25f_m' A_n + 0.65A_{st}F_s)$$

$$\times \left(1 - \left(\frac{h}{140r}\right)^2\right) \quad [h/r < 99] \qquad 69.26$$

$$P_a = (0.25f_m' A_n + 0.65A_{st}F_s)\left(\frac{70r}{h}\right)^2$$

$$[h/r > 99] \qquad 69.27$$

$$k_b = \frac{F_b}{F_b + \dfrac{F_s}{n}} \qquad 69.28$$

11. ASD WALL DESIGN: SHEAR WALLS WITH NO NET TENSION—UNREINFORCED

$$f_v = \frac{VQ}{Ib} = \frac{3V}{2A_n} \quad \text{[rectangular sections]} \qquad 69.29$$

F_v is the least of

(a) $1.5\sqrt{f'_m}$

(b) 120 lbf/in^2 (0.83 MPa)

(c) $v + 0.45 N_v/A_n$

12. ASD WALL DESIGN: SHEAR WALLS WITH NET TENSION—REINFORCED

$$f_v = \frac{V}{bd} \qquad\qquad 69.30$$

$$F_v = \frac{1}{3}\left(4 - \frac{M}{Vd}\right)\sqrt{f'_m}$$
$$< 80 - 45\left(\frac{M}{Vd}\right) \quad [M/Vd < 1] \qquad 69.31$$

$$F_v = \sqrt{f'_m}$$
$$< 35 \text{ lbf/in}^2 \quad (0.24 \text{ MPa}) \quad [M/Vd \geq 1] \quad 69.32$$

$$F_v = \frac{1}{2}\left(4 - \frac{M}{Vd}\right)\sqrt{f'_m}$$
$$< 120 - 45\left(\frac{M}{Vd}\right) \quad [M/Vd < 1] \qquad 69.33$$

$$F_v = 1.5\sqrt{f'_m}$$
$$< 75 \text{ lbf/in}^2 \quad (0.52 \text{ MPa}) \quad [M/Vd \geq 1] \quad 69.34$$

$$A_v = \frac{Vs}{F_s d} \qquad\qquad 69.35$$

CERM Chapter 70
Masonry Columns

> Chapter, section, equation, figure, and table numbers correspond to CERM. For additional study material, go to the corresponding chapter and section number in CERM.

5. DESIGNING FOR PURE COMPRESSION

$$P_a = (0.25 f'_m A_n + 0.65 A_{st} F_s)$$
$$\times \left(1 - \left(\frac{h}{140r}\right)^2\right) \quad [h/r < 99] \qquad 70.1$$

$$P_a = (0.25 f'_m A_n + 0.65 A_{st} F_s)$$
$$\times \left(\frac{70r}{h}\right)^2 \quad [h/r > 99] \qquad 70.2$$

7. BIAXIAL BENDING

$$\frac{1}{P_{\text{biaxial}}} = \frac{1}{P_x} + \frac{1}{P_y} - \frac{1}{P_o} \quad [P > 0.1 P_o] \qquad 70.3$$

Transportation

CERM Chapter 71
Properties of Solid Bodies

Chapter, section, equation, figure, and table numbers correspond to CERM. For additional study material, go to the corresponding chapter and section number in CERM.

1. CENTER OF GRAVITY

$$x_c = \frac{\int x \, dm}{m} \qquad 71.1$$

$$y_c = \frac{\int y \, dm}{m} \qquad 71.2$$

$$z_c = \frac{\int z \, dm}{m} \qquad 71.3$$

$$x_c = \frac{\sum m_i x_{ci}}{\sum m_i} \qquad 71.4$$

$$y_c = \frac{\sum m_i y_{ci}}{\sum m_i} \qquad 71.5$$

$$z_c = \frac{\sum m_i z_{ci}}{\sum m_i} \qquad 71.6$$

2. MASS AND WEIGHT

$$m = \rho V \qquad 71.7$$

$$w = \frac{mg}{g_c} \qquad 71.8(b)$$

4. MASS MOMENT OF INERTIA

$$I_x = \int (y^2 + z^2) \, dm \qquad 71.9$$

$$I_y = \int (x^2 + z^2) \, dm \qquad 71.10$$

$$I_z = \int (x^2 + y^2) \, dm \qquad 71.11$$

5. PARALLEL AXIS THEOREM

$$I_{\text{any parallel axis}} = I_c + md^2 \qquad 71.12$$

$$I = I_{c,1} + m_1 d_1^2 + I_{c,2} + m_2 d_2^2 + \cdots \quad 71.13$$

6. RADIUS OF GYRATION

$$k = \sqrt{\frac{I}{m}} \qquad 71.14$$

$$I = k^2 m \qquad 71.15$$

CERM Chapter 72
Kinematics

Chapter, section, equation, figure, and table numbers correspond to CERM. For additional study material, go to the corresponding chapter and section number in CERM.

5. LINEAR PARTICLE MOTION

$$s(t) = \int v(t) \, dt = \int \left(\int a(t) \, dt \right) dt \qquad 72.2$$

$$v(t) = \frac{ds(t)}{dt} = \int a(t) \, dt \qquad 72.3$$

$$a(t) = \frac{dv(t)}{dt} = \frac{d^2 s(t)}{dt^2} \qquad 72.4$$

$$v_{\text{ave}} = \frac{\int_1^2 v(t) \, dt}{t_2 - t_1} = \frac{s_2 - s_1}{t_2 - t_1} \qquad 72.5$$

$$a_{\text{ave}} = \frac{\int_1^2 a(t) \, dt}{t_2 - t_1} = \frac{v_2 - v_1}{t_2 - t_1} \qquad 72.6$$

6. DISTANCE AND SPEED

$$\text{displacement} = s(t_2) - s(t_1) \qquad 72.7$$

7. UNIFORM MOTION

$$s(t) = s_0 + \mathrm{v}t \qquad \text{72.8}$$

$$\mathrm{v}(t) = \mathrm{v} \qquad \text{72.9}$$

$$a(t) = 0 \qquad \text{72.10}$$

8. UNIFORM ACCELERATION

$$a(t) = a \qquad \text{72.11}$$

$$\mathrm{v}(t) = a \int dt = \mathrm{v}_0 + at \qquad \text{72.12}$$

$$s(t) = a \iint dt^2 = s_0 + \mathrm{v}_0 t + \tfrac{1}{2}at^2 \qquad \text{72.13}$$

Table 72.1 Uniform Acceleration Formulas[a]

to find	given these	use this equation
a	$t, \mathrm{v}_0, \mathrm{v}$	$a = \dfrac{\mathrm{v} - \mathrm{v}_0}{t}$
a	t, v_0, s	$a = \dfrac{2s - 2\mathrm{v}_0 t}{t^2}$
a	$\mathrm{v}_0, \mathrm{v}, s$	$a = \dfrac{\mathrm{v}^2 - \mathrm{v}_0^2}{2s}$
s	t, a, v_0	$s = \mathrm{v}_0 t + \tfrac{1}{2}at^2$
s	$a, \mathrm{v}_0, \mathrm{v}$	$s = \dfrac{\mathrm{v}^2 - \mathrm{v}_0^2}{2a}$
s	$t, \mathrm{v}_0, \mathrm{v}$	$s = \tfrac{1}{2}t(\mathrm{v}_0 + \mathrm{v})$
t	$a, \mathrm{v}_0, \mathrm{v}$	$t = \dfrac{\mathrm{v} - \mathrm{v}_0}{a}$
t	a, v_0, s	$t = \dfrac{\sqrt{\mathrm{v}_0^2 + 2as} - \mathrm{v}_0}{a}$
t	$\mathrm{v}_0, \mathrm{v}, s$	$t = \dfrac{2s}{\mathrm{v}_0 + \mathrm{v}}$
v_0	t, a, v	$\mathrm{v}_0 = \mathrm{v} - at$
v_0	t, a, s	$\mathrm{v}_0 = \dfrac{s}{t} - \tfrac{1}{2}at$
v_0	a, v, s	$\mathrm{v}_0 = \sqrt{\mathrm{v}^2 - 2as}$
v	t, a, v_0	$\mathrm{v} = \mathrm{v}_0 + at$
v	a, v_0, s	$\mathrm{v} = \sqrt{\mathrm{v}_0^2 + 2as}$

[a] The table can be used for rotational problems by substituting α, ω, and θ for a, v, and s, respectively.

CERM Chapter 73
Kinetics

> Chapter, section, equation, figure, and table numbers correspond to CERM. For additional study material, go to the corresponding chapter and section number in CERM.

5. LINEAR MOMENTUM

$$\mathbf{p} = \frac{m\mathbf{v}}{g_c} \qquad \text{73.1(b)}$$

9. NEWTON'S SECOND LAW OF MOTION

$$F = \left(\frac{m}{g_c}\right)\left(\frac{d\mathrm{v}}{dt}\right) = \frac{ma}{g_c} \qquad \text{73.12(b)}$$

$$T = \left(\frac{I}{g_c}\right)\left(\frac{d\omega}{dt}\right) = \frac{I\alpha}{g_c} \qquad \text{73.14(b)}$$

10. CENTRIPETAL FORCE

$$F_c = \frac{m\mathrm{v}_t^2}{g_c r} \qquad \text{73.15(b)}$$

13. FLAT FRICTION

$$N = \frac{mg}{g_c} \qquad \text{73.19(b)}$$

$$N = \frac{mg\cos\phi}{g_c} \qquad \text{73.20(b)}$$

$$F_{f,\max} = f_s N \qquad \text{73.21}$$

$$\tan\phi = f_s \qquad \text{73.22}$$

16. ROLLING RESISTANCE

$$F_r = \frac{mga}{rg_c} = \frac{wa}{r} \qquad \text{73.27(b)}$$

$$f_r = \frac{F_r}{w} = \frac{a}{r} \qquad \text{73.28}$$

17. MOTION OF RIGID BODIES

$$\sum F_y = ma_y \quad \text{[consistent units]} \qquad \text{73.30}$$

$$T = \frac{I\alpha}{g_c} \qquad \text{73.31(b)}$$

20. IMPULSE

$$\text{Imp} = \int_{t_1}^{t_2} F \, dt \quad \text{[linear]} \qquad 73.44$$

$$\text{Imp} = \int_{t_1}^{t_2} T \, dt \quad \text{[angular]} \qquad 73.45$$

$$\text{Imp} = F(t_2 - t_1) \quad \text{[linear]} \qquad 73.46$$

$$\text{Imp} = T(t_2 - t_1) \quad \text{[angular]} \qquad 73.47$$

21. IMPULSE-MOMENTUM PRINCIPLE

$$F(t_2 - t_1) = \frac{m(\text{v}_2 - \text{v}_1)}{g_c} \qquad 73.50(b)$$

$$T(t_2 - t_1) = \frac{I(\omega_2 - \omega_1)}{g_c} \qquad 73.51(b)$$

22. IMPULSE-MOMENTUM PRINCIPLE IN OPEN SYSTEMS

$$F = \frac{m \Delta \text{v}}{g_c \Delta t} = \frac{\dot{m} \Delta \text{v}}{g_c} \qquad 73.52(b)$$

23. IMPACTS

$$m_1 \text{v}_1 + m_2 \text{v}_2 = m_1 \text{v}_1' + m_2 \text{v}_2' \qquad 73.53$$

Figure 73.17 *Direct Central Impact*

$$\left. m_1 \text{v}_1^2 + m_2 \text{v}_2^2 = m_1 \text{v}_1'^2 + m_2 \text{v}_2'^2 \right|_{\text{elastic impact}} \qquad 73.54$$

24. COEFFICIENT OF RESTITUTION

$$e = \frac{\text{relative separation velocity}}{\text{relative approach velocity}}$$

$$= \frac{\text{v}_1' - \text{v}_2'}{\text{v}_2 - \text{v}_1} \qquad 73.55$$

33. ROADWAY BANKING

$$\tan \phi = \frac{\text{v}_t^2}{gr} \qquad \text{[no friction]} \qquad 73.90$$

$$\text{v}_t = \sqrt{gr \tan \phi} \qquad 73.91$$

$$e = \tan \phi = \frac{\text{v}_t^2 - fgr}{gr + f\text{v}_t^2} \quad \text{[with friction]} \qquad 73.92$$

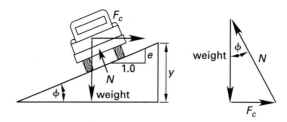

Figure 73.24 *Roadway Banking*

CERM Chapter 74
Roads and Highways: Capacity Analysis

Chapter, section, equation, figure, and table numbers correspond to CERM. For additional study material, go to the corresponding chapter and section number in CERM.

8. VOLUME PARAMETERS

$$K = \frac{\text{DHV}}{\text{AADT}} \qquad 74.1$$

$$\text{DDHV} = D(\text{DHV}) = DK(\text{AADT}) \qquad 74.2$$

$$\text{volume-capacity ratio}_i = (v/c)_i \qquad 74.3$$

$$v_{m,i} = c \, (v/c)_{m,i} \qquad 74.4$$

$$\text{PHF} = \frac{\text{actual hourly volume}_{\text{vph}}}{\text{peak rate of flow}_{\text{vph}}}$$

$$= \frac{V_{\text{vph}}}{v_p} = \frac{V_{\text{vph}}}{4V_{15 \text{ min,peak}}} \qquad 74.5$$

$$V_i = (v_{m,i})N \left(\frac{\text{adjustment}}{\text{factors}} \right) \qquad 74.6$$

9. TRIP GENERATION

$$\ln (\text{no. of trips}) = A + B \log \left(\frac{\text{calling population}}{\text{parameter}} \right) \qquad 74.7$$

$$\text{no. of trips} = \frac{C}{\left(\dfrac{\text{calling population}}{\text{parameter}} \right)^D} \qquad 74.8$$

10. SPEED, FLOW, AND DENSITY RELATIONSHIPS

$$S = S_b \left(1 - \frac{D}{D_j}\right) \qquad \text{74.9}$$

$$v = SD = \frac{3600 \, \frac{\text{sec}}{\text{hr}}}{\text{headway} \, \frac{\text{sec}}{\text{veh}}} \qquad \text{74.10}$$

$$\text{spacing}_{\text{ft/veh}} = \frac{5280 \, \frac{\text{ft}}{\text{mi}}}{D_{\text{vpm/lane}}} \qquad \text{74.11(b)}$$

$$\text{headway}_{\text{sec/veh}} = \frac{\text{spacing}_{\text{ft/veh}}}{\text{space mean speed}_{\text{ft/sec}}} \qquad \text{74.12(b)}$$

$$v_{\text{vph}} = \frac{3600 \, \frac{\text{sec}}{\text{hr}}}{\text{headway}_{\text{sec/veh}}} \qquad \text{74.13}$$

13. FREEWAYS

$$v = c\,(v/c) \qquad \text{74.15}$$

$$v_{m,i} = c\,(v/c)_{m,i} \qquad \text{74.16}$$

$$V = v_p(\text{PHF})Nf_{\text{HV}}f_p \qquad \text{74.17}$$

$$f_{\text{HV}} = \frac{1}{1 + P_T(E_T - 1) + P_R(E_R - 1)} \qquad \text{74.18}$$

$$\text{FFS} = \text{BFFS} - f_{\text{LW}} - f_{\text{LC}} - f_N - f_{\text{ID}} \qquad \text{74.19}$$

14. MULTILANE HIGHWAYS

$$\text{FFS} = \text{BFFS} - f_M - f_{\text{LW}} - f_{\text{LC}} - f_A \qquad \text{74.20}$$

$$v_p = D_m(\text{FFS}) = \frac{V}{N(\text{PHF})f_{\text{HV}}f_P} \qquad \text{74.21}$$

15. SIGNALIZED INTERSECTIONS

$$c_i = s_i \left(\frac{g_i}{C}\right) \qquad \text{74.22}$$

$$s_{\text{vphgpl}} = \frac{3600 \, \frac{\text{sec}}{\text{hr}}}{\text{saturation headway}_{\text{sec/veh}}} \qquad \text{74.23}$$

$$X_i = \left(\frac{v}{c}\right)_i = \frac{v_i}{s_i\left(\frac{g_i}{C}\right)} = \frac{v_i C}{s_i g_i} \qquad \text{74.24}$$

$$X_c = \sum_i \left(\frac{v}{s}\right)_{ci} \left(\frac{C}{C - L}\right) \qquad \text{74.25}$$

$$s = s_o N f_w f_{\text{HV}} f_g f_p f_{\text{bb}} f_a f_{\text{LU}} f_{\text{RT}} f_{\text{LT}} f_{\text{Lpb}} f_{\text{Rpb}} \qquad \text{74.26}$$

19. ON-DEMAND TIMING

$$\text{no. of cars in initial period} = \frac{\text{distance between line and detector}}{\text{car length}} \qquad \text{74.28}$$

20. WALKWAYS

$$v = \frac{v_{p,15}}{15W_E} \qquad \text{74.29}$$

$$v = SD = \frac{S}{M} \qquad \text{74.30}$$

28. QUEUING MODELS

$$L = \lambda W \qquad \text{74.31}$$

$$L_q = \lambda W_q \qquad \text{74.32}$$

$$W = W_q + \frac{1}{\mu} \qquad \text{74.33}$$

29. M/M/1 SINGLE-SERVER MODEL

$$f(t) = \mu e^{-\mu t} \qquad \text{74.34}$$

$$P\{t > h\} = e^{-\mu h} \qquad \text{74.35}$$

$$p\{x\} = \frac{e^{-\lambda}\lambda^x}{x!} \qquad \text{74.36}$$

$$p\{0\} = 1 - \rho \qquad \text{74.37}$$

$$p\{n\} = p\{0\}\rho^n \qquad \text{74.38}$$

$$W = \frac{1}{\mu - \lambda} = W_q + \frac{1}{\mu} = \frac{L}{\lambda} \qquad \text{74.39}$$

$$W_q = \frac{\rho}{\mu - \lambda} = \frac{L_q}{\lambda} \qquad \text{74.40}$$

$$L = \frac{\lambda}{\mu - \lambda} = L_q + \rho \qquad \text{74.41}$$

$$L_q = \frac{\rho\lambda}{\mu - \lambda} \qquad \text{74.42}$$

30. M/M/s MULTI-SERVER MODEL

$$W = W_q + \frac{1}{\mu} \qquad 74.43$$

$$W_q = \frac{L_q}{\lambda} \qquad 74.44$$

$$L_q = \frac{p\{0\}\rho\left(\frac{\lambda}{\mu}\right)^s}{s!(1-\rho)^2} \qquad 74.45$$

$$L = L_q + \frac{\lambda}{\mu} \qquad 74.46$$

$$p\{0\} = \cfrac{1}{\cfrac{\left(\frac{\lambda}{\mu}\right)^s}{s!\left(1-\frac{\lambda}{s\mu}\right)} + \sum_{j=0}^{s-1}\cfrac{\left(\frac{\lambda}{\mu}\right)^j}{j!}} \qquad 74.47$$

$$p\{n\} = \frac{p\{0\}\left(\frac{\lambda}{\mu}\right)^n}{n!} \quad [n \le s] \qquad 74.48$$

$$p\{n\} = \frac{p\{0\}\left(\frac{\lambda}{\mu}\right)^n}{s!s^{n-s}} \quad [n > s] \qquad 74.49$$

32. DETOURS

$$L_{ft} = \frac{W_{ft}S_{mph}^2}{60} \quad [\text{v} \le 40 \text{ mph; U.S. only}] \qquad 74.50(a)$$

$$L_{ft} = W_{ft}S_{mph} \quad [\text{v} > 40 \text{ mph; U.S. only}] \qquad 74.51(a)$$

CERM Chapter 75
Vehicle Dynamics and Accident Analysis

Chapter, section, equation, figure, and table numbers correspond to CERM. For additional study material, go to the corresponding chapter and section number in CERM.

1. VEHICLE DYNAMICS

$$F_i = \frac{ma}{g_c} = \frac{wa}{g} \qquad 75.1(b)$$

$$F_g = w\sin\phi$$

$$\approx w\tan\phi = \frac{wG\%}{100} \qquad 75.2$$

$$F_r = f_r w\cos\phi$$

$$\approx f_r w \qquad 75.3$$

$$F_D = \frac{C_D A\rho \text{v}^2}{2g_c} \qquad 75.4(b)$$

$$F_D = KA\text{v}^2 \approx 0.0006 A_{ft^2}\text{v}_{mi/hr}^2 \qquad 75.5(b)$$

$$P = (F_i + F_g + F_r + F_c + F_D)\text{v} \qquad 75.6$$

$$\text{v}_{vehicle} = \text{v}_{t,tire} = \pi d_{tire}n_{tire} \qquad 75.7$$

$$n_{wheel} = \frac{n_{engine}}{R_{transmission}R_{differential}} \qquad 75.8$$

$$R_{transmission} = \frac{n_{engine}}{n_{driveshaft}} \qquad 75.9$$

$$R_{differential} = \frac{n_{driveshaft}}{n_{wheel}} \qquad 75.10$$

$$P_{hp} = \frac{T_{in\text{-}lbf}n_{rpm}}{63,025} \qquad 75.11(b)$$

$$F_{tractive} = \frac{\eta_m T R_{transmission}R_{differential}}{r_{tire}} \qquad 75.12$$

$$\text{v} = \frac{2\pi n_{rev/sec}(1-i)}{R_{transmission}R_{differential}} \qquad 75.13$$

$$\dot{m}_{fuel,lbm/hr} = P_{brake,hp}(BSFC_{lbm/hp\text{-}hr}) \qquad 75.14(b)$$

$$\dot{Q}_{fuel,gal/hr} = \frac{\dot{m}_{fuel,lbm/hr}\left(7.48\,\dfrac{gal}{ft^3}\right)}{\rho_{lbm/ft^3}} \qquad 75.15(b)$$

$$s'_{fuel,mi/gal} = \frac{\text{v}_{mi/hr}}{\dot{Q}_{fuel,gal/hr}} \qquad 75.16(b)$$

2. DYNAMICS OF STEEL-WHEELED RAILROAD ROLLING STOCK

$$\begin{aligned}DBP &= TFDA - LR \\ &= CR + AF \end{aligned} \qquad 75.17$$

$$F_{tractive,lbf} = \frac{375P_{hp,rated}\eta_{drive}}{\text{v}_{mph}} \qquad 75.18(b)$$

$$\begin{aligned}R_{lbf/ton} &= 0.6 + \frac{20}{w_{tons}} \\ &+ 0.01\text{v}_{mph} + \frac{K\text{v}_{mph}^2}{w_{tons}n}\end{aligned} \qquad 75.19$$

3. COEFFICIENT OF FRICTION

$$f = \frac{F_f}{N} \qquad 75.20$$

5. STOPPING DISTANCE

$$s_{stopping} = \text{v}t_p + s_b \qquad 75.21$$

6. BRAKING AND DECELERATION RATE

$$a = fg = f\left(32.2\ \frac{\text{ft}}{\text{sec}^2}\right) \qquad \text{75.22(b)}$$

7. BRAKING AND SKIDDING DISTANCE

$$s_b = \frac{\text{v}^2}{2g(f\cos\theta + \sin\theta)} \qquad \text{75.23}$$

$$s_b = \frac{\text{v}^2}{2g(f + G)}$$

$$\approx \frac{\text{v}_{\text{mph}}^2}{30(f + G)} \qquad \text{75.24(b)}$$

9. ANALYSIS OF ACCIDENT DATA

$$R_{\text{RMEV}} = \frac{(\text{no. of accidents})(10^6)}{(\text{ADT})(\text{no. of years})\left(365\ \dfrac{\text{days}}{\text{yr}}\right)} \qquad \text{75.25}$$

$$R_{\text{HMVM}} = \frac{(\text{no. of accidents})(10^8)}{(\text{ADT})(\text{no. of years})\left(365\ \dfrac{\text{days}}{\text{yr}}\right)L_{\text{mi}}}$$

$$\text{75.26}$$

CERM Chapter 76
Flexible Pavement Design

> Chapter, section, equation, figure, and table numbers correspond to CERM. For additional study material, go to the corresponding chapter and section number in CERM.

9. WEIGHT-VOLUME RELATIONSHIPS

$$
\begin{aligned}
P_b = &(100\%)(\text{aggregate surface area}) \\
&\times (\text{asphalt thickness}) \\
&\times (\text{specific weight of asphalt}) \qquad \text{76.1}
\end{aligned}
$$

10. PLACEMENT AND PAVING EQUIPMENT

$$L_{\text{ft/ton}} = \frac{2000\ \dfrac{\text{lbf}}{\text{ton}}}{w_{\text{ft}}t_{\text{ft}}\gamma_{\text{compacted,lbf/ft}^3}} \qquad \text{76.2(b)}$$

$$L_{\text{ft/ton}} = \frac{18{,}000\ \dfrac{\text{lbf-ft}^2}{\text{ton-yd}^2}}{r_{\text{lbf/yd}^2}w_{\text{ft}}} \qquad \text{76.3(b)}$$

$$
\begin{aligned}
\text{v}_{\text{ft/min}} &= \frac{R_{p,\text{ton/hr}}L_{\text{ft/ton}}}{60\ \dfrac{\text{min}}{\text{hr}}} \\[2mm]
&= \frac{R_{p,\text{ton/hr}}\left(2000\ \dfrac{\text{lbf}}{\text{ton}}\right)\left(12\ \dfrac{\text{in}}{\text{ft}}\right)}{w_{\text{ft}}t_{\text{in}}\gamma_{\text{lbf/ft}^3}\left(60\ \dfrac{\text{min}}{\text{hr}}\right)}
\end{aligned}
$$

$$\text{76.4(b)}$$

12. CHARACTERISTICS OF ASPHALT CONCRETE

$$G_{\text{sa}} = \frac{W}{V_{\text{aggregate}}\gamma_{\text{water}}} \qquad \text{76.5(b)}$$

$$G_{\text{sb}} = \frac{P_1 + P_2 + \ldots P_n}{\dfrac{P_1}{G_1} + \dfrac{P_2}{G_2} + \ldots + \dfrac{P_n}{G_n}} \qquad \text{76.6}$$

$$G_{\text{sa}} = \frac{A}{A - C} \qquad \text{76.7}$$

$$G_{\text{sb}} = \frac{A}{B - C} \qquad \text{76.8}$$

$$\text{absorption} = \frac{(100\%)(B - A)}{A} \qquad \text{76.9}$$

$$G_{\text{sa}} = \frac{A}{B + A - C} \qquad \text{76.10}$$

$$G_{\text{sb}} = \frac{A}{B + S - C} \qquad \text{76.11}$$

$$\text{absorption} = \frac{(100\%)(S - A)}{A} \qquad \text{76.12}$$

$$G_{\text{se}} = \frac{100\% - P_b}{\dfrac{100\%}{G_{\text{mm}}} - \dfrac{P_b}{G_b}} \qquad \text{76.13}$$

$$G_{\text{mm}} = \frac{100\%}{\dfrac{P_s}{G_{\text{se}}} + \dfrac{P_b}{G_b}} \qquad \text{76.14}$$

$$G_{\text{mm}} = \frac{A}{A + D - E} \qquad \text{76.15}$$

$$P_{\text{ba}} = \frac{(100\%)G_b(G_{\text{se}} - G_{\text{sb}})}{G_{\text{sb}}G_{\text{se}}} \qquad \text{76.16}$$

$$P_{be} = P_b - \frac{P_{ba}P_s}{100\%} \qquad 76.17$$

$$\text{VMA} = 100\% - \frac{G_{mb}P_s}{G_{sb}} \qquad 76.18$$

$$P_a = \text{VTM} = \frac{(100\%)(G_{mm} - G_{mb})}{G_{mm}} \qquad 76.19$$

$$\text{VFA} = \frac{(100\%)(\text{VMA} - \text{VTM})}{\text{VMA}} \qquad 76.20$$

13. MARSHALL MIX TEST PROCEDURE

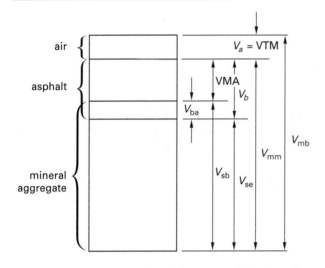

VMA = volume of voids in mineral aggregate
V_{mb} = bulk volume of compacted mix
V_{mm} = voidless volume of paving mix
V_a = volume of air voids
V_b = volume of asphalt
V_{ba} = volume of absorbed asphalt
V_{sb} = volume of mineral aggregate (by bulk specific gravity)
V_{se} = volume of mineral aggregate (by effective specific gravity)

Figure 76.2 *Volumes in a Compacted Asphalt Specimen*
(see also Fig. 76.1)

15. TRUCK FACTORS

$$\text{TF} = \frac{\text{ESALs}}{\text{no. of trucks}} \qquad 76.22$$

16. DESIGN TRAFFIC

$$\text{ESAL}_{20} = (\text{ESAL}_{\text{first year}})(\text{GF}) \qquad 76.23$$

$$w_{18} = D_D D_L \hat{w}_{18} \qquad 76.24$$

21. PAVEMENT STRUCTURAL NUMBER

$$\text{SN} = D_1 a_1 + D_2 a_2 m_2 + D_3 a_3 m_3 \qquad 76.29$$

$$a_2 = 0.249(\log_{10} E_b) - 0.977$$
$$[\textit{AASHTO Guide} \text{ p. II-20}] \qquad 76.30$$

$$a_3 = 0.227(\log_{10} E_{sb}) - 0.839$$
$$[\textit{AASHTO Guide} \text{ p. II-22}] \qquad 76.31$$

CERM Chapter 77
Rigid Pavement Design

Chapter, section, equation, figure, and table numbers correspond to CERM. For additional study material, go to the corresponding chapter and section number in CERM.

4. MATERIAL STRENGTHS

$$k = \frac{M_R}{19.4} \qquad 77.1$$

$$E_c = 57{,}000\sqrt{f'_c} \quad [E_c \text{ and } f'_c \text{ in lbf/in}^2] \qquad 77.2$$

8. STEEL REINFORCING

$$P_s = \frac{L_{\text{ft}}F(100\%)}{2f_{s,\text{lbf/in}^2}} \qquad 77.3$$

$$P_t = \frac{A_s(100\%)}{YD} \qquad 77.4$$

CERM Chapter 78
Plane Surveying

Chapter, section, equation, figure, and table numbers correspond to CERM. For additional study material, go to the corresponding chapter and section number in CERM.

1. ERROR ANALYSIS: MEASUREMENTS OF EQUAL WEIGHT

$$x_p = \frac{x_1 + x_2 + \cdots + x_k}{k} \qquad 78.1$$

$$E_{\text{mean}} = \frac{0.6745s}{\sqrt{k}}$$

$$= \frac{E_{\text{total},k \text{ measurements}}}{\sqrt{k}} \qquad 78.2$$

2. ERROR ANALYSIS: MEASUREMENTS OF UNEQUAL WEIGHT

$$E_{p,\text{weighted}} = 0.6745\sqrt{\frac{\sum\left(w_i(\bar{x} - x_i)^2\right)}{(k-1)\sum w_i}} \qquad 78.3$$

3. ERRORS IN COMPUTED QUANTITIES

$$E_{\text{sum}} = \sqrt{E_1^2 + E_2^2 + E_3^2 + \cdots} \qquad 78.4$$

$$E_{\text{product}} = \sqrt{x_1^2 E_2^2 + x_2^2 E_1^2} \qquad 78.5$$

13. DISTANCE MEASUREMENT: TAPING

$$C_T = L\alpha(T - T_s) \quad \text{[temperature]} \qquad 78.6$$

$$C_P = \frac{(P - P_s)L}{AE} \quad \text{[tension]} \qquad 78.7$$

$$C_s = \pm\left(\frac{W^2 L^3}{24P^2}\right) \quad \text{[sag]} \qquad 78.8$$

14. DISTANCE MEASUREMENT: TACHYOMETRY

$$x = K(R_2 - R_1) + C \qquad 78.9$$

$$x = K(R_2 - R_1)\cos^2\theta + C\cos\theta \qquad 78.10$$

$$y = \tfrac{1}{2}K(R_2 - R_1)\sin 2\theta + C\sin\theta \qquad 78.11$$

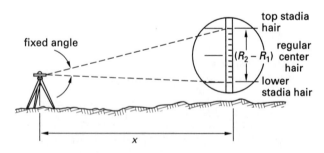

Figure 78.1 Horizontal Stadia Measurement

18. ELEVATION MEASUREMENT

$$h_c = \left(2.4 \times 10^{-8}\,\frac{1}{\text{ft}}\right)x_{\text{ft}}^2 \qquad 78.12$$

$$h_r = \left(3.0 \times 10^{-9}\,\frac{1}{\text{ft}}\right)x_{\text{ft}}^2 \qquad 78.13$$

$$h_a = R_{\text{observed}} + h_r - h_c$$
$$= R_{\text{observed}} - \left(2.1 \times 10^{-8}\,\frac{1}{\text{ft}}\right)x_{\text{ft}}^2 \qquad 78.14$$

19. ELEVATION MEASUREMENT: DIRECT LEVELING

$$y_{\text{A-L}} = R_A - h_{\text{rc,A-L}} - \text{HI} \qquad 78.15$$

$$y_{\text{L-B}} = \text{HI} + h_{\text{rc,L-B}} - R_B \qquad 78.16$$

$$y_{\text{A-B}} = y_{\text{A-L}} + y_{\text{L-B}}$$
$$= R_A - R_B + h_{\text{rc,L-B}} - h_{\text{rc,A-L}} \qquad 78.17$$

$$y_{\text{A-B}} = R_A - R_B \qquad 78.18$$

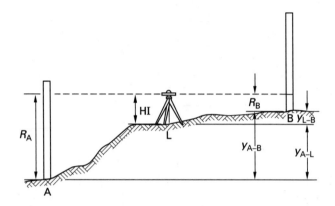

Figure 78.4 Direct Leveling

21. ELEVATION MEASUREMENT: INDIRECT LEVELING

$$y_{\text{A-B}} = \text{AC}\tan\alpha + 2.1 \times 10^{-8}(\text{AC})^2 \qquad 78.19$$

$$y_{\text{A-B}} = \text{AC}\tan\beta - 2.1 \times 10^{-8}(\text{AC})^2 \qquad 78.20$$

$$y_{\text{A-B}} = \tfrac{1}{2}\text{AC}(\tan\alpha + \tan\beta) \qquad 78.21$$

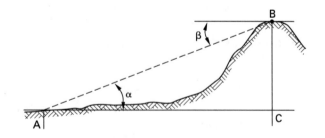

Figure 78.5 Indirect Leveling

23. DIRECTION SPECIFICATION

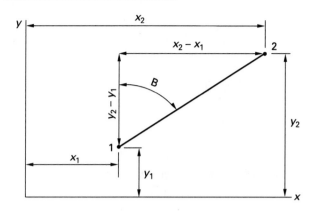

Figure 78.6 *Calculation of Bearing Angle*

$$\tan B = \frac{x_2 - x_1}{y_2 - y_1} \qquad 78.22$$

$$D = \sqrt{(y_2 - y_1)^2 + (x_2 - x_1)^2} \qquad 78.23$$

$$\tan A_N = \frac{x_2 - x_1}{y_2 - y_1} \qquad 78.24$$

$$\tan A_S = \frac{x_1 - x_2}{y_1 - y_2} \qquad 78.25$$

25. TRAVERSES

1. The sum of the deflection angles is 360°.
2. The sum of the interior angles of a polygon with n sides is $(n-2)(180°)$.

28. BALANCING CLOSED TRAVERSE DISTANCES

$$L = \sqrt{\begin{array}{c}(\text{closure in departure})^2 \\ + (\text{closure in latitude})^2\end{array}} \qquad 78.28$$

$$\frac{\text{leg departure correction}}{\text{closure in departure}} = \frac{-\text{leg length}}{\text{total traverse length}} \qquad 78.29$$

$$\frac{\text{leg latitude correction}}{\text{closure in latitude}} = \frac{-\text{leg length}}{\text{total traverse length}} \qquad 78.30$$

$$\frac{\text{leg departure correction}}{\text{closure in departure}}$$
$$= -\left(\frac{\text{leg departure}}{\text{sum of departure absolute values}}\right) \qquad 78.31$$

$$\frac{\text{leg latitude correction}}{\text{closure in latitude}}$$
$$= -\left(\frac{\text{leg latitude}}{\text{sum of latitude absolute values}}\right) \qquad 78.32$$

30. TRAVERSE AREA: METHOD OF COORDINATES

$$A = \tfrac{1}{2}\left|\left(\sum_{i=1}^{n} y_i(x_{i-1} - x_{i+1})\right)\right| \qquad 78.33$$

$$\frac{x_1}{y_1} \times \frac{x_2}{y_2} \times \frac{x_3}{y_3} \times \frac{x_4}{y_4} \times \frac{x_1}{y_1} \qquad \text{etc.}$$

$$A = \tfrac{1}{2}\,|\textstyle\sum \text{ of full line products}$$
$$-\textstyle\sum \text{ of broken line products}| \qquad 78.34$$

31. TRAVERSE AREA: DOUBLE MERIDIAN DISTANCE

$$\text{DMD}_{\text{leg } i} = \text{DMD}_{\text{leg } i\text{-}1} + \text{departure}_{\text{leg } i\text{-}1}$$
$$+ \text{departure}_{\text{leg } i} \qquad 78.35$$

$$A = \tfrac{1}{2}\,|\textstyle\sum(\text{latitude}_{\text{leg } i} \times \text{DMD}_{\text{leg } i})| \qquad 78.36$$

32. AREAS BOUNDED BY IRREGULAR BOUNDARIES

$$A = d\left(\frac{h_1 + h_n}{2} + \sum_{i=2}^{n-1} h_i\right) \qquad 78.37$$

$$A = \left(\frac{d}{3}\right) h_1 - h_n + 2 \underset{\substack{\text{for all odd numbers} \\ \text{starting at } i=3}}{\sum^{n-1} h_i} + 4 \underset{\substack{\text{for all even numbers} \\ \text{starting at } i=2}}{\sum^{n-1} h_i}$$
$$78.38$$

33. PHOTOGRAMMETRY

$$\text{scale} = \frac{\text{focal length}}{\text{flight altitude}}$$
$$= \frac{\text{length in photograph}}{\text{true length}} \qquad 78.39$$

CERM Chapter 79
Horizontal, Compound, Vertical, and Spiral Curves

Chapter, section, equation, figure, and table numbers correspond to CERM. For additional study material, go to the corresponding chapter and section number in CERM.

1. HORIZONTAL CURVES

$$R = \frac{5729.578}{D} \quad \text{[U.S.—arc definition]} \quad \textbf{79.1}$$

$$R = \frac{50}{\sin\left(\dfrac{D}{2}\right)} \quad \text{[U.S.—chord definition]} \quad \textbf{79.2}$$

$$L = \frac{2\pi R I}{360°} = R I_{\text{radians}} = 100\left(\frac{I}{D}\right) \quad \textbf{79.3}$$

$$T = R\tan\left(\frac{I}{2}\right) \quad \textbf{79.4}$$

$$E = R\left(\sec\left(\frac{I}{2}\right) - 1\right) = R\tan\left(\frac{I}{2}\right)\tan\left(\frac{I}{4}\right) \quad \textbf{79.5}$$

$$M = R\left(1 - \cos\left(\frac{I}{2}\right)\right) = \left(\frac{C}{2}\right)\tan\left(\frac{I}{4}\right) \quad \textbf{79.6}$$

$$C = 2R\sin\left(\frac{I}{2}\right) = 2T\cos\left(\frac{I}{2}\right) \quad \textbf{79.7}$$

2. DEGREE OF CURVE

$$D = \frac{(360°)(100)}{2\pi R} = \frac{5729.578}{R} \quad \text{[arc basis]} \quad \textbf{79.8}$$

$$\sin\left(\frac{D}{2}\right) = \frac{50}{R} \quad \text{[chord basis]} \quad \textbf{79.9}$$

$$L \approx \left(\frac{I}{D}\right)(100 \text{ ft}) \quad \textbf{79.10}$$

3. STATIONING ON A HORIZONTAL CURVE

$$\text{sta PT} = \text{sta PC} + L \quad \textbf{79.11}$$
$$\text{sta PC} = \text{sta PI} - T \quad \textbf{79.12}$$

4. CURVE LAYOUT BY DEFLECTION ANGLE

(1) The deflection angle between a tangent and a chord (Fig. 79.3(a)) is half of the arc's subtended angle.

(2) The angle between two chords (Fig. 79.3(b)) is half of the arc's subtended angle.

$$\alpha = \angle\text{V-PC-A} = \frac{\beta}{2} \quad \textbf{79.13}$$

$$\frac{\beta}{360°} = \frac{\text{arc length PC-A}}{2\pi R} \quad \textbf{79.14}$$

$$\frac{\beta}{I} = \frac{\text{arc length PC-A}}{L} \quad \textbf{79.15}$$

$$C_{\text{PC-A}} = 2R\sin\alpha = 2R\sin\left(\frac{\beta}{2}\right) \quad \textbf{79.16}$$

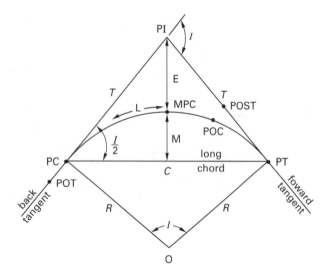

Figure 79.1 *Horizontal Curve Elements*

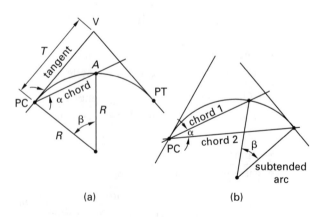

(a)   (b)

Figure 79.3 *Circular Curve Deflection Angle*

5. TANGENT OFFSETS

$$y = R(1 - \cos\beta)$$
$$= R - \sqrt{R^2 - x^2} \qquad \textit{79.17}$$

$$\beta = \arcsin\left(\frac{x}{R}\right) = \arccos\left(\frac{R - y}{R}\right) \qquad \textit{79.18}$$

$$x = R\sin\beta = \sqrt{2Ry - y^2} \qquad \textit{79.19}$$

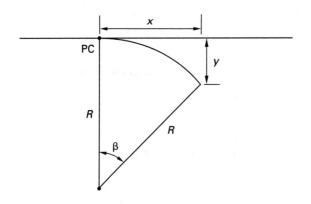

Figure 79.4 *Tangent Offset*

6. CURVE LAYOUT BY TANGENT OFFSETS

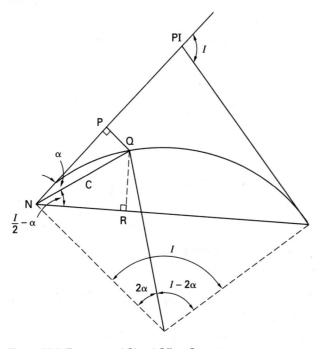

Figure 79.5 *Tangent and Chord Offset Geometry*

$$NP = \text{tangent distance} = NQ\cos\alpha \qquad \textit{79.21}$$

$$PQ = \text{tangent offset} = NQ\sin\alpha \qquad \textit{79.22}$$

$$NQ = C = 2R\sin\alpha \qquad \textit{79.23}$$

$$NP = (2R\sin\alpha)\cos\alpha$$
$$= C\cos\alpha \qquad \textit{79.24}$$

$$PQ = (2R\sin\alpha)\sin\alpha$$
$$= 2R\sin^2\alpha \qquad \textit{79.25}$$

7. CURVE LAYOUT BY CHORD OFFSET

$$NR = \text{chord distance} = NQ\cos\left(\frac{I}{2} - \alpha\right)$$
$$= (2R\sin\alpha)\cos\left(\frac{I}{2} - \alpha\right)$$
$$= C\cos\left(\frac{I}{2} - \alpha\right) \qquad \textit{79.26}$$

$$RQ = \text{chord offset} = NQ\sin\left(\frac{I}{2} - \alpha\right)$$
$$= (2R\sin\alpha)\sin\left(\frac{I}{2} - \alpha\right)$$
$$= C\sin\left(\frac{I}{2} - \alpha\right) \qquad \textit{79.27}$$

8. HORIZONTAL CURVES THROUGH POINTS

$$\alpha = \arctan\left(\frac{y}{x}\right) \qquad \textit{79.28}$$

$$m = \sqrt{x^2 + y^2} \qquad \textit{79.29}$$

$$\gamma = 90° - \frac{I}{2} - \alpha \qquad \textit{79.30}$$

$$\phi = 180° - \arcsin\left(\frac{\sin\gamma}{\cos\left(\frac{I}{2}\right)}\right) \qquad \textit{79.31}$$

$$\theta = 180° - \gamma - \phi \qquad \textit{79.32}$$

$$\frac{\sin\theta}{m} = \frac{\sin\phi\cos\left(\frac{I}{2}\right)}{R} \qquad \textit{79.33}$$

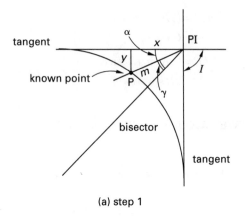

(a) step 1

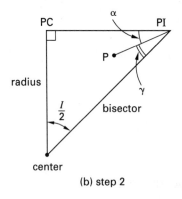

(b) step 2

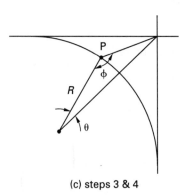

(c) steps 3 & 4

Figure 79.6 *Horizontal Curve Through a Point (I known)*

10. SUPERELEVATION

$$F_c = \frac{mv_t^2}{R} \quad \text{[consistent units]} \qquad 79.34$$

$$e = \tan\phi = \frac{v^2}{gR} \quad \text{[consistent units]} \qquad 79.35$$

$$e = \tan\phi = \frac{v^2}{gR} - f_s \quad \text{[consistent units]} \qquad 79.36$$

$$e = \tan\phi = \frac{v_{mph}^2}{15R} - f_s \qquad 79.37(b)$$

$$f_s = 0.16 - \frac{0.01(v_{mph} - 30)}{10} \quad [< 50 \text{ mph}] \qquad 79.38$$

$$f_s = 0.14 - \frac{0.02(v_{mph} - 50)}{10} \quad [50 \text{ to } 70 \text{ mph}] \qquad 79.39$$

11. TRANSITIONS TO SUPERELEVATION

$$T_R = \frac{wp}{SRR} \qquad 79.40$$

$$L = \frac{we}{SRR} \qquad 79.41$$

12. SUPERELEVATION OF RAILROAD LINES

$$E = \frac{G_{eff}v^2}{gR} \quad \text{[railroads]} \qquad 79.42$$

13. STOPPING SIGHT DISTANCE

$$S = \left(1.47 \frac{\frac{ft}{sec}}{\frac{mi}{hr}}\right) t_p v_{mph} + \frac{v_{mph}^2}{30(f + G)} \qquad 79.43(b)$$

15. MINIMUM HORIZONTAL CURVE LENGTH FOR STOPPING DISTANCE

$$S = \left(\frac{R}{28.65}\right)\left(\arccos\left(\frac{R - M}{R}\right)\right) \qquad 79.44$$

$$M = R(1 - \cos\theta) = R\left(1 - \cos\left(\frac{DS}{200}\right)\right) \qquad 79.45$$

$$= R\left(1 - \cos\left(\frac{28.65S}{R}\right)\right) \qquad 79.45$$

16. VERTICAL CURVES

$$R = \frac{G_2 - G_1}{L} \quad \text{[may be negative]} \qquad 79.46$$

$$elev_x = \left(\frac{R}{2}\right)x^2 + G_1 x + elev_{BVC} \qquad 79.47$$

$$x_{turning\ point} = \frac{-G_1}{R} \quad \text{[in stations]} \qquad 79.48$$

$$M_{ft} = \frac{AL_{sta}}{8} \qquad 79.49$$

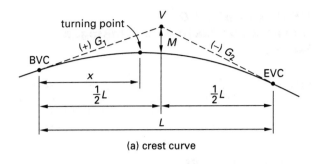

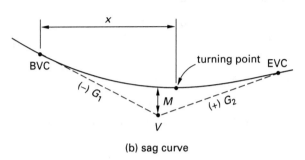

Figure 79.9 *Symmetrical Parabolic Vertical Curve*

17. VERTICAL CURVES THROUGH POINTS

$$s = \sqrt{\frac{\text{elev}_E - \text{elev}_G}{\text{elev}_E - \text{elev}_F}} \qquad 79.50$$

$$L = \frac{2d(s+1)}{s-1} \qquad 79.51$$

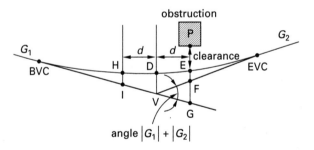

Figure 79.10 *Vertical Curve with an Obstruction*

18. VERTICAL CURVE TO PASS THROUGH TURNING POINT

$$L = \frac{2(\text{elev}_V - \text{elev}_{TP})}{G_1\left(\dfrac{G_1}{G_2 - G_1} + 1\right)} \qquad 79.52$$

19. MINIMUM VERTICAL CURVE LENGTH FOR SIGHT DISTANCES (CREST CURVES)

$$L = \frac{AS^2}{200\left(\sqrt{h_1} + \sqrt{h_2}\right)^2} \quad [S < L] \qquad 79.53$$

$$L = 2S - \frac{200\left(\sqrt{h_1} + \sqrt{h_2}\right)^2}{A} \quad [S > L] \qquad 79.54$$

$$L = \frac{AS^2}{800(C-5)} \quad [S < L] \qquad 79.55(b)$$

$$L = 2S - \frac{800(C-5)}{A} \quad [S > L] \qquad 79.56(b)$$

Table 79.4 *AASHTO Required Lengths of Curves on Grades*[a]

	stopping sight distance[b] (crest curves)	passing sight distance (crest curves)	stopping sight distance (sag curves)
$S < L$	$\dfrac{AS^2}{1329}$	$\dfrac{AS^2}{3093}$	$\dfrac{AS^2}{400 + 3.5S}$
$S > L$	$2S - \dfrac{1329}{A}$	$2S - \dfrac{3093}{A}$	$2S - \dfrac{400 + 3.5S}{A}$

[a] $A = G_1 - G_2$, the algebraic difference in grades.
[b] The driver's eye is 3.50 ft (1070 mm) above road surface, viewing an object 0.5 ft (150 mm) high.

Compiled from *A Policy on Geometric Design of Highways and Streets*, Chapter III, copyright © 1990 by the American Association of State Highway and Transportation Officials, Washington, D.C. Used by permission.

20. DESIGN OF CREST CURVES USING *K*-VALUE

$$K = \frac{L}{A} = \frac{L}{|G_2 - G_1|} \quad [\text{always positive}] \qquad 79.57$$

22. MINIMUM VERTICAL CURVE LENGTH FOR COMFORT: SAG CURVES

$$L_{\text{ft}} = \frac{A\text{v}_{\text{mph}}^2}{46.5} \qquad 79.58(b)$$

25. SPIRAL CURVES

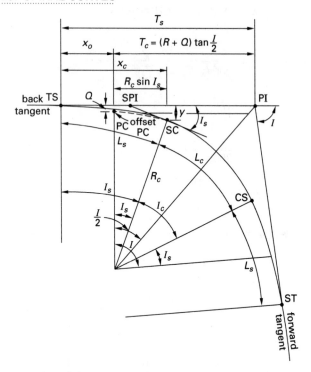

Figure 79.16 *Spiral Curve Geometry*

$$L_{s,\text{ft}} \approx \frac{1.6 \text{v}_{\text{mph}}^3}{R_{\text{ft}}} \qquad 79.62(b)$$

$$I_s = \left(\frac{L_s}{100}\right)\left(\frac{D}{2}\right) = \frac{L_s D}{200} \qquad 79.63$$

$$I = I_c + 2I_s \qquad 79.64$$

$$L = L_c + 2L_s \qquad 79.65$$

$$\alpha_s = \tan^{-1}\frac{y}{x} = \frac{y}{x} \approx \frac{I_s}{3} \qquad 79.66$$

$$\frac{\alpha_P}{\alpha_s} = \frac{I_P}{I_s} = \left(\frac{L_P}{L_s}\right)^2 \qquad 79.67$$

$$y = x \tan\alpha_p \approx x\alpha_{p,\text{radians}}$$

$$= \frac{xI_{s,\text{radians}}}{3} = \left(\frac{xL_s D_{\text{degrees}}}{(200)(3)}\right)\left(\frac{\pi}{180°}\right)$$

$$= \frac{xL_s}{6R} \quad [I_s < 20°] \qquad 79.68$$

$$T_c = (R_c + Q)\tan\frac{I}{2} \qquad 79.69$$

$$Q = y - R_c(1 - \cos I_s)$$

$$= \frac{L_s^2}{6R} - R_c(1 - \cos I_s) \qquad 79.70$$

CERM Chapter 80
Construction Earthwork, Staking, and Surveying

Chapter, section, equation, figure, and table numbers correspond to CERM. For additional study material, go to the corresponding chapter and section number in CERM.

1. VOLUMES OF PILES

$$V = \left(\frac{h}{3}\right)\pi r^2 \quad [\text{cone}] \qquad 80.1$$

$$V = \left(\frac{h}{6}\right)b(2a + a_1) = \tfrac{1}{6}hb\left(3a - \frac{2h}{\tan\phi}\right)[\text{wedge}] \quad 80.2$$

$$V = \left(\frac{h}{6}\right)\left(ab + (a + a_1)(b + b_1) + a_1 b_1\right)$$

$$= \left(\frac{h}{6}\right)\left(ab + 4\left(a - \frac{h}{\tan\phi_1}\right)\left(b - \frac{h}{\tan\phi_2}\right)\right.$$

$$\left. + \left(a - \frac{2h}{\tan\phi_1}\right)\left(b - \frac{2h}{\tan\phi_2}\right)\right)$$

$$\left[\begin{array}{c}\text{frustrum of a}\\\text{rectangular pyramid}\end{array}\right] \qquad 80.3$$

$$a_1 = a - \frac{2h}{\tan\phi_1} \qquad 80.4$$

$$b_1 = b - \frac{2h}{\tan\phi_2} \qquad 80.5$$

(a) cone

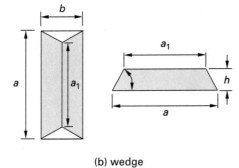

(b) wedge

(c) frustrum of a rectangular pyramid

Figure 80.1 *Pile Shapes*

3. AVERAGE END AREA METHOD

$$V = \frac{L(A_1 + A_2)}{2} \qquad 80.6$$

$$V_{\text{pyramid}} = \frac{LA_{\text{base}}}{3} \qquad 80.7$$

4. PRISMOIDAL FORMULA METHOD

$$V = \left(\frac{L}{6}\right)(A_1 + 4A_m + A_2) \qquad 80.8$$

Systems, Management, and Professional

CERM Chapter 81
Project Management

Chapter, section, equation, figure, and table numbers correspond to CERM. For additional study material, go to the corresponding chapter and section number in CERM.

4. SOLVING A CPM PROBLEM

ES	EF
LS	LF

key

ES: Earliest Start

EF: Earliest Finish

LS: Latest Start

LF: Latest Finish

step 1: Place the project start time or date in the **ES** and **EF** positions of the start activity. The start time is zero for relative calculations.

step 2: Consider any unmarked activity, all of whose predecessors have been marked in the **EF** and **ES** positions. (Go to step 4 if there are none.) Mark in its **ES** position the largest number marked in the **EF** position of those predecessors.

step 3: Add the activity time to the **ES** time and write this in the **EF** box. Go to step 2.

step 4: Place the value of the latest finish date in the **LS** and **LF** boxes of the finish mode.

step 5: Consider unmarked predecessors whose successors have all been marked. Their **LF** is the smallest **LS** of the successors. Go to step 7 if there are no unmarked predecessors.

step 6: The **LS** for the new node is **LF** minus its activity time. Go to step 5.

step 7: The slack time for each node is **LS−ES** or **LF−EF**.

step 8: The critical path encompasses nodes for which the slack time equals **LS−ES** from the start node. There may be more than one critical path.

6. STOCHASTIC CRITICAL PATH MODELS

$$\mu = \tfrac{1}{6}(t_{\text{minimum}} + 4t_{\text{most likely}} + t_{\text{maximum}}) \qquad 81.1$$

$$\sigma = \tfrac{1}{6}(t_{\text{maximum}} - t_{\text{minimum}}) \qquad 81.2$$

$$p\{\text{duration} > D\} = p\{t > z\} \qquad 81.3$$

$$z = \frac{D - \mu_{\text{critical path}}}{\sigma_{\text{critical path}}} \qquad 81.4$$

CERM Chapter 85
Engineering Economic Analysis

Chapter, section, equation, figure, and table numbers correspond to CERM. For additional study material, go to the corresponding chapter and section number in CERM.

11. SINGLE-PAYMENT EQUIVALENCE

$$F = P(1+i)^n \qquad 85.2$$

$$P = F(1+i)^{-n} = \frac{F}{(1+i)^n} \qquad 85.3$$

12. STANDARD CASH FLOW FACTORS AND SYMBOLS

Table 85.1 Discount Factors for Discrete Compounding

factor name	converts	symbol	formula
single payment compound amount	P to F	$(F/P, i\%, n)$	$(1+i)^n$
single payment present worth	F to P	$(P/F, i\%, n)$	$(1+i)^{-n}$
uniform series sinking fund	F to A	$(A/F, i\%, n)$	$\dfrac{i}{(1+i)^n - 1}$
capital recovery	P to A	$(A/P, i\%, n)$	$\dfrac{i(1+i)^n}{(1+i)^n - 1}$
uniform series compound amount	A to F	$(F/A, i\%, n)$	$\dfrac{(1+i)^n - 1}{i}$
uniform series present worth	A to P	$(P/A, i\%, n)$	$\dfrac{(1+i)^n - 1}{i(1+i)^n}$
uniform gradient present worth	G to P	$(P/G, i\%, n)$	$\dfrac{(1+i)^n - 1}{i^2(1+i)^n} - \dfrac{n}{i(1+i)^n}$
uniform gradient future worth	G to F	$(F/G, i\%, n)$	$\dfrac{(1+i)^n - 1}{i^2} - \dfrac{n}{i}$
uniform gradient uniform series	G to A	$(A/G, i\%, n)$	$\dfrac{1}{i} - \dfrac{n}{(1+i)^n - 1}$

24. CHOICE OF ALTERNATIVES: COMPARING ONE ALTERNATIVE WITH ANOTHER ALTERNATIVE

Capitalized Cost Method

$$\text{capitalized cost} = \text{initial cost} + \frac{\text{annual costs}}{i} \qquad 85.19$$

$$\text{capitalized cost} = \text{initial cost} + \frac{\text{EAA}}{i}$$

$$= \text{initial cost} + \frac{\text{present worth}}{\text{of all expenses}} \qquad 85.20$$

25. CHOICE OF ALTERNATIVES: COMPARING AN ALTERNATIVE WITH A STANDARD

Benefit-Cost Ratio Method

$$B/C = \frac{\Delta \substack{\text{user} \\ \text{benefits}}}{\Delta \substack{\text{investment} \\ \text{cost}} + \Delta\text{maintenance} - \Delta \substack{\text{residual} \\ \text{value}}} \qquad 85.21$$

26. RANKING MUTUALLY EXCLUSIVE MULTIPLE PROJECTS

$$\frac{B_2 - B_1}{C_2 - C_1} \geq 1 \quad [\text{alternative 2 superior}] \qquad 85.22$$

36. DEPRECIATION METHODS
Straight Line Method

$$D = \frac{C - S_n}{n}$$ *85.25*

Statutory Depreciation Systems

$$D_j = C \times \text{factor}$$ *85.33*

38. BOOK VALUE

$$\text{BV}_j = C - \sum_{m=1}^{j} D_m$$ *85.43*

41. BASIC INCOME TAX CONSIDERATIONS

$$t = s + f - sf$$ *85.44*

45. RATE AND PERIOD CHANGES

$$\phi = \frac{r}{k}$$ *85.53*

$$i = (1 + \phi)^k - 1$$

$$= \left(1 + \frac{r}{k}\right)^k - 1$$ *85.54*

53. BREAK-EVEN ANALYSIS

$$C = f + aQ$$ *85.58*

$$R = pQ$$ *85.59*

$$Q^* = \frac{f}{p - a}$$ *85.60*

56. INFLATION

$$i' = i + e + ie$$ *85.62*

Index